Rationale-Based Software Engineering

Janet E. Burge · John M. Carroll ·
Raymond McCall · Ivan Mistrik

Rationale-Based
Software Engineering

 Springer

Authors

Janet E. Burge
Miami University
School of Eng. & Appl. Science
Computer Science & Systems Analysis
205 Benton Hall
Oxford, OH 45056
USA
burgeje@muohio.edu

John M. Carroll
Penn State University
School of Information Sciences
and Technology
504 Rider I Building
120 S. Burrowes Street
University Park, PA 16801-3857
USA
jcarroll@ist.psu.edu

Raymond McCall
University of Colorado
College Architecture & Planning
314 UCB
Boulder, CO 80309-0314
USA
Mccall@colorado.edu

Ivan Mistrík
Independent Consultant
Werderstr. 45
69120 Heidelberg
Germany
i.j.mistrik@t-online.de

ISBN 978-3-642-09631-0   e-ISBN 978-3-540-77583-6

ACM Computing Classification (1998): D.2, K.6

© 2010 Springer-Verlag Berlin Heidelberg

Cover design: KünkelLopka Werbeagentur, Heidelberg, Germany

Printed on acid-free paper

9 8 7 6 5 4 3 2 1

springer.com

Foreword

The Search for Meaning

At the risk of appearing to exaggerate, I will argue that the pursuit of rationale in engineering is nothing less than a *search for meaning*. On the face of it, capturing, recording, and perusing rationale in support of software engineering is a worthy software management activity, whose benefits are well documented and accepted. Indeed chapters of this book speak to this issue. However, there is a more significant reason for the pursuit of rationale: it is a desire to make sense of the world – to explain it and to explain its behavior, both expected and unexpected. Weick calls this *sense-making*, and of course is right insofar as the world makes 'sense'. Wouldn't it be grand if we are able to understand why the world is structured as it is, and why artifacts in the world have been engineered to behave the way they do? Sometimes, the reasons why are straight forward: an engineer solving a problem in the world may recognize it as a *normal problem* that he has encountered before, the solution of which is well understood, tried, and tested. The rationale for his engineering solution in this case is mostly reusable – afterall he is engaging in normal engineering, in *normal design* (Vincenti 1990).

But what if the problem encountered is *radical?* Well, Vincenti tells us that we need to engage in radical engineering, in *radical design*. The consequence of this is that we should expect to fail in our first attempts at a solution, but strive to learn from our failures, so that future encounters with our radical problem become more normal.

It is in this transition from radical to normal that rationale research offers attractive opportunities for advancing the state-of-the-art in software engineering, and offers an intellectual umbrella for breaking new ground in this area. This umbrella needs to cover both *problem analysis* in pursuit of stakeholder requirements, and *engineering design* in pursuit of solutions to those requirements. Research on the relationship between requirements and design, on managing traceability and software evolution, and, ultimately, on assuring the quality of software engineering solutions, all sit comfortably under this umbrella.

However, there is a difficulty, observed by Jackson (Jackson 2007). As we specialize and strive to evolve the discipline of software engineering into normal engineering, we find that much normal design rationale is hidden, perhaps lost in time, when imaginable alternative solutions were considered and discarded. We then find ourselves back to where we started: trying to make sense of what we have already, trying to understand the reasons why the normal design we have before us is the way it is. Weick writes:

> Sensemaking is about such things as placement of items into frameworks, comprehending, redressing surprise, constructing meaning, interacting in pursuit of mutual understanding, and patterning. (Weick 1995, p. 6)

My colleague Simon Buckingham Shum (Buckingham Shum 2007) has taken Weick's message to heart, and has made sensemaking the centre piece of his framework for constructive argumentation and explanatory rationale.

A book such as this is important because the development of software is *engineering* and not science. It is not enough simply to understand why software behaves the way it does, but rather how it can be built – rationally (Parnas and Clements 1986), or at least systematically – to behave as intended. We need the framework offered by this volume to develop such meaningful software.

Bashar Nuseibeh
Professor of Computing and Director of Research
The Open University, UK

Foreword

Design Rationale: Retrospect and Prospect

Danish philosopher Soren Kierkegaard once said that life can only be understood backwards but must be lived forward. He might easily have been talking about the software design process. In a software project, many developers work together on a system development effort, some of them only for some phases of the project, and few with an overview of the entire system. As a system emerges from this process, it has to be explained for future designers, maintenance programmers and others. Some of the logic, the rationale for the way it the system is may have been apparent from the beginning. But in many cases, the trajectory of all the little decisions that contribute to the way a system is when it is released may only be apparent in retrospect when tied together in a deliberate activity of sense-making and documentation.

It is not supposed to be this way. Software engineering textbooks and process improvement manuals exhort us to have well defined requirements and a core architectural vision that drive the details forward. Such requirements and architectural features must be documented and internalized by project staff so that everyone appreciates the significance and impact of changes. But the real life of software projects is not so simple. Stakeholders change their minds. The business context of a system changes during its development. As infrastructure technology changes, new implementation opportunities become available. Architectural commitments have to be changed or diluted as their consequences become apparent. People simply forget what they are doing and develop different styles in how they work and implement software that may conflict. Kierkegaard, of course, was not a software architect. His struggle was not with customers, users and intransigent or imperfect fellow developers, but nineteenth-century institutions, repression and hypocrisy. But the consequences of both struggles are the same: we cannot always make sense of what is going on when we are in the middle of things until after the smoke has cleared. And the root causes are essentially the same, too: the world is complex, and people are only human.

Design rationale research started in the 1980s from the recognition that the results of design often do not make sense to those outside the design team, but have to be made sense of to foster better understanding during the ongoing process of maintenance and feature evolution. There were two strands to this research emerging from different communities, and these strands persist to the present day. In a nutshell, the difference between them goes to Kierkegaard's comment: if we have to make a choice between the two alternative modes of sense making, should we try to make design easier to explain in retrospect, or should we make it more transparent and reflective while it is going on? These answers led to retrospective rationale research and prospective rationale research.

In the retrospective rationale community, the concern was primarily how to document large-scale software architectures or standards. Since an architecture is a stable foundation for a continually evolving system, and standards are expected to endure over generations of many systems produced by many organizations, it is essential to document the architecture and standards and the reasons behind them. Architectures and standards are the types of thing that we are stuck with once we make a commitment to them and then come to depend on them for a myriad of detailed decisions. They are therefore high-risk commitments.

The first notable example of the use of rationale in after-the-fact documentation came from the team led by David Parnas, who used the avionics software of the A-7 aircraft as a microcosm for exploring design specification and documentation techniques (Parnas and Clements 1986).

The A-7 work has had an influence in more recent efforts and methods, such as the SEI architecture initiative (Clements et al. 2003). Different but similarly motivated efforts have led to architectural decision support technology such as ADSS (Capilla et al. 2007). There is a general consensus emerging that the documentation of rationale, whatever form it takes, should be tightly bound to the documentation about the architecture itself. For example, Zhu and Gorton (2007) devised a UML profile for adding rationale information to standard UML design diagrams. In this way, the rationale is a first-class part of the documentation, not an addendum or collection of low-value notes.

In the standards community, it has become almost universal to document the rationale for parts of a standard, often in terms of comparisons between what the standard requires and plausible but inferior alternatives. A pioneering example of this in software engineering was the documentation and rationale for the Ada programming language and programming support environment. The latter (code named "Stoneman") is a particularly good example of the analysis of why the environment had a layered structure (Druffel and Buxton 1980).

The explanatory role of retrospective rationale documentation means that it is not critical that the rationale be historically accurate. Designers might well have had considerations in mind that led to architecture decisions that no longer seem relevant in retrospect. Conversely, making sense of the overall sweep of the architectural design process in retrospect, it may be clear that reasons were prominent that were not completely apparent at the time. Reasons given may be a white lie, they may oversimplify or distort that convoluted process that people went through. They may even be self-serving or apologetic, designed to protect the authors from criticism. After the fact, these factors may not be important to maintenance programmers or designers of later releases. An approximate, simplified and glossed rationale, even one that is somewhat distorted, may be more suitable for supporting the concerns of these professionals than documentation that more faithfully describes the agonizing process of decision making that the developers actually went through.

In keeping with this creative and constructive distortion of the design decision making process is another property of retrospective rationale: it has to be carefully crafted. Like the documentation of the architecture's form, the documentation of its rationale is expected to endure and become part of the project's knowledge base as the project goes forward. It therefore makes economic sense to invest resources and time in writing rationale documents and clearly articulating their audiences.

In contrast, another community came to design rationale research in the 1980s. These researchers had been inspired by the pioneering work of Horst Rittel and other design theorists in their attempts to provide structure to collaborative decision making among designers and other stakeholders, particularly in community architecture and urban planning projects, when the problems they faced were "wicked". A wicked problem is characterized by dispute about what the problem actually is and how one would recognize whether it had been solved. Thus, a wicked problem is not just hard; its very nature is contested and cries out for discussion. In this second tradition of design rationale research, therefore, an emphasis was placed on semi-structured representations of ongoing issues, positions and arguments. The emphasis was on supporting problem formulation and decision making as they occurred rather than seeking to justify decisions for people who came after.

Such support is support for design rationale, though, for two reasons. The first reason is definitional: The rationale for a decision consists of the reasons why it was chosen. These reasons do not have to be documented with a future consumer, such as a maintenance programmer, in mind. Even if nobody were to read the rationale in the future, its documentation and value during the unfolding of the design would not make it any the less a record of rationale. A second reason for this type of ongoing decision making support counting as rationale is more practical: The information may well be useful in the future by accident, even though that may not be the motivation for its capture.

In fact, the term "capture" reveals a fundamental difference between the two streams of research. Both streams emphasize care and professionalism during design. But the careful audience analysis, crafting and writing of rationale documentation in the retrospective rationale tradition emphasizes that the recording of rationale is a significant part of a project and should be budgeted for and rewarded. In contrast, the capturing of design rationale in the prospective rationale tradition implies that rationale documentation is a fortuitously gathered by-product of another activity. That other activity, collaborative design argumentation and decision making, may be serious, it may be planned and budgeted, and it may be highly structured in its processes. But the rationale produced is expected to be immediately valuable, and any later benefits that accrue from it should not require any further planning or writing. These benefits should come for free.

Probably the most influential prototype prospective rationale management system was gIBIS (Conklin and Begeman 1989), which although it never created a major user community, it was used in NCR for the development of hotel and restaurant support systems (Conklin and Burgess-Yakemovic 1991), and the IBIS argumentation model at its core has been extremely influential as the baseline for representation of nearly all rationale.

In parallel to the use of rationale capture in software engineering, a similar argumentation model based on explicit decision criteria was influencing research into user interface design in the human-computer interaction (HCI) community (Maclean et al. 1989). Here the design decisions were typically more local in scope, such as in the choice of alternative user interface widgets or menu structures to support a user's task. The model used, Design Space Analysis, based on questions, options, and criteria, rather than the issues, positions and arguments of IBIS, emphasized the making of choices between mutually exclusive options and was based on explicit and frequently quantitative criteria. The design problems addressed were therefore constrained and clearly specified. While they may have been subtle and far reaching in their impact on usability, they were anything but wicked problems in the sense defined above.

More recent research in prospective design rationale in software engineering tends to emphasize quantitative criteria for choosing among alternatives, and the normal targets of these decisions are architectural choices such as the distribution of services across a network. Recent work in software engineering economics represents an attempt to make these design decisions rigorous in the same way that financial decisions in business can be based on rigorous projections and risk models (Boehm et al. 1995; Bose 1998). Typical of the decision-making methods and models that are incorporated into such work are Cost-Benefit Analysis (e.g. Kazman et al. 2003) and the Analytical Hierarchy Process (e.g. Lozano-Tello and Gomez-Perez 2001; Wallin et al. 2007).

In software engineering, other than the early, limited experiments with gIBIS, prospective design rationale research foundered for several years, possibly because the unique qualities of software design were largely neglected. The increasingly complex argumentation models could have applied to the design of anything. In customizing design rationale representations to software engineering, a key early insight was that software design methods are sets of heuristics for making requirements, design and implementation decisions, not just software notations. Object-oriented methods, for example, provide guidelines for the identification of objects and their res-ponsibilities and guidelines for refactoring when these early decisions lead to reorganization. Such methods essentially encapsulate reusable design knowledge. Before the widespread adoption of object-oriented methods, the methodology community was rather fragmented, and so extensions of design rationale representations to software methods tended to focus on illustrative methods. Among these were Potts and Bruns's (1988) adaptation of the Liskov and Guttag abstraction-based design method, the Potts (1989) treatment of Jackson's JSD, and the later incorporation of goal-based and scenario-based representations of system requirements into the Inquiry Cycle model of prospective rationale (Potts et al. 1994).

In addition to design methods, which tend to focus on generic design decisions, domain-specific issues can also arise that can be captured and represented as reusable rationale. An early example of this was Belotti's (1993) attempt to integrate theoretical and practitioner perspectives on HCI guidelines through Design Space Analysis. More recent work has included domain-specific architectural rationale for automotive software engineering (Wallin et al. 2007).

Not satisfied with the attachment of design rationale to design artifacts represented by these methodological extensions or by the introduction of explicit and often quantitative criteria in the Design Space Analysis community in HCI, some researchers sought to extend the rationale models in such a way that some decisions could be made automatically or dependencies between decisions could be computed and maintained consistently (Ali Babar and Gorton, 2007; Lee, 1991; Lee and Lai 1991; Wang and Xiong

2001) by means of an elaborate and formal data model for design rationale information and its relationship to elements in the design itself. It is not clear to what extent the benefits of such computational support outweigh the burden of recording the rationale information in such a rigorous and necessarily fine-grained fashion. Nor is it clear whether the structure of such models can be easily maintained as the design and its rationale change. Such approaches do promise to be extremely valuable when coupled with a formal theory of design change and configuration management.

Recently, however, the focus has returned to reusable software engineering knowledge and rationale. The entire software patterns community (Gamma et al. 1995) can be regarded as engaged in a quest to produce a corpus of rationale documents that discuss design patterns, the issues that arise when they are used, the arguments for when they are appropriate and when they cause problems, how they interact, and illustrations of their use. Debating whether a pattern library is a generic library of design artifacts or a rationale library seems rather fruitless, since the design alternatives faced by a designer, criteria and considerations that affect the decisions, illustrative solutions, and warnings about interactions are so inextricably interwoven. The role of rationale is woven through the patterns literature, although it has a secondary role to the capturing of artifact knowledge. A more explicit role for rationale can be seen in Baniassad et al.'s (2003) Design Pattern Rationale Graph (DPRG), a tool for linking designs and implementations through rationale information.

Thus, much of the development of work in the prospective design rationale tradition can be seen to be aimed at producing information that can be used subsequently, not just as an aid to decision making in the moment. Such subsequent uses may include specific rationale information to be referred to later in the same project, or it may even take the form of more generic lessons learned that can be applied across projects.

There have been few comprehensive reviews on design rationale, and none of monograph length. The theoretical models of design rationale, the phases of the design process during which they are useful, the domain-specific contexts in which they can be applied, and evidence of practicality have been lacking. Only one software engineering textbook, that of Bruegge and Dutoit (2004) makes a thoroughgoing attempt to integrate design rationale into the software engineering process.

As we enter the third decade of design rationale research, however, now is a good time to take stock. Everyone acknowledges that designing is difficult, that it involves many people often over long periods of time who need explicit records of who did what, when and for what reasons. The support for design rationale and its integration into software engineering processses has not yet reached the mainstream of software engineering writing and practice, and it is time that it did. Or as Kierkegaard also said: Truth always rests with the minority.

Colin Potts
Associate Professor in the School of Interactive Computing
Georgia Institute of Technology, USA

Preface

The most distinctive thing about humans is not the thumb, of course. It's design. Unlike any other animal, we incessantly and dramatically reshape both ourselves and our environment. We design ourselves through innovating concepts, language, culture and other practices, and we design almost everything around us. It is telling that we now speak of "natural" places on the Earth to distinguish the few places we have not (yet!) redesigned.

Among the most complex, diverse, and pervasive things that humans design is software systems. The history of software design is almost entirely a history of trying to catch up with complexity and diversity. As we look back to the 1960s the notion of what was then called the "software crisis" seems almost amusing. At that time, barely a decade after the invention of software, it was recognized that the complexity and diversity of software systems was being elaborated far more rapidly than were engineering techniques to manage software development. What is amusing is that this was (optimistically) called a *crisis*, as if it were a temporary threat that would in the course of time be rectified.

But this never happened. Instead the software crisis became chronic. It became the context for the software industry and for software engineering. And by now, as almost every system is, incorporates, or fundamentally depends upon software, as software systems have become utterly pervasive, the software crisis has really become an epoch in human history.

No one is very happy about this, and from time to time manifestations of the on-going software crisis bubble up into dramatic mass media reports about how vital defense systems are fundamentally unverifiable, about how medical systems make it more or less inevitable than surgeons will kill their patients, about how banking systems occasionally share account information with unknown hackers, and so forth.

What are we to do? There are many answers, many approaches, but none of them is a "silver bullet" (as Fred Brooks vividly put this). The most obvious approach, and quite likely the most powerful, is to explicitly describe and justify the design, implementation, and use of software systems, and to do this routinely, iteratively, and regularly throughout the software development process. We call this "Rationale-Based Software

Engineering." It is not a new idea, though in some areas there are new tools and techniques. Rather, it is an essential idea that has been around, that we cannot afford to lose track of, and that perhaps can be pushed to a greater fruition now. In this book, we try to bring together a broad discussion of rationale and focus it on aspects of the very old and very weighty challenge of the software crisis.

Book Overview

This book consists of four parts. Part 1 sets the context for the work and describes why Software Engineering Rationale (SER) and Rationale-Based Software Engineering (RBSE) are essential contributors toward improving the software development process. Part 2 describes how Software Engineering Rationale can be used to support software development. Part 3 describes how RBSE can be applied throughout the software engineering lifecycle as well as supporting software reuse. Part 4 presents architectural and conceptual frameworks for RBSE as well as our vision of future directions for RBSE research.

Part 1: Introduction

So why capture rationale? Before making a case for why SER capture and use should be an essential part of software development, it is important to first define what it is. Part 1 defines rationale and sets the context for the remainder of the book.

Chapter 1, "What is Rationale and Why Does it Matter" provides an initial discussion of the scope and value of rationale in software engineering. An initial introduction of previous work on rationale is provided and we make our initial case for why rationale is useful during software engineering.

Chapter 2, "What Makes Software Different" describes some of the key differences between applying rationale to software engineering and applying rationale to other domains. This includes both opportunities for use in software engineering that are lacking when developing other artifacts as well as some of the unique challenges posed by software development. Specifically, we look at the role of the computer in software development vs. physical artifact development as well as the implications of the necessity to support iteration in software development on rationale management.

Chapter 3, "Rationale and Software Engineering" introduces both Software Engineering and Software Engineering Rationale (Dutoit et al. 2006).

Rationale has a role to play in defining software processes, supporting software project management, and as a mechanism to both document and guide decision-making throughout the software process.

Chapter 4, "Learning from Rationale Research in Other Domains" describes key rationale research in other domains and its implication to software engineering. The chapter focuses on four areas: domain-oriented design environments using PHI (McCall 1991); automating design rationale capture in Computer Aided Design, more specifically that using the Rationale Construction Framework (Myers et al. 1999); rationale support via Parameter Dependency Networks and DRIVE (De la Garza and Alcantara 1997); and how Case Based Reasoning (CBR) systems such ARCHIE (Zimring et al. 1995) relate to rationale.

Chapter 5, "Decision-making in Software Engineering" examines the role that human decision-making has in software engineering. The chapter describes naturalistic decision-making and Klein's recognition-primed decision model (Klein 1998), which addresses some of the problems with classical decision making by proposing a strategy more consistent with observations of human decision-makers, where the first acceptable alternative is selected. The chapter concludes with a discussion of rationale as a resource for decision-making and how rationale relates to both the classical and naturalistic views.

Part 2: Uses of Rationale

There is little or no point in capturing rationale if there are not ways in which it can be used. Part 2 describes some key uses of rationale in software development.

Chapter 6, "Presentation of Rationale" looks at rationale presentation. The two major classes of presentation formats, semi-formal and informal, are described. The chapter then describes new opportunities for presentation provided by reusable rationale databases, multi-scale presentation, and development tool integration.

Chapter 7, "Evaluation" describes how rationale can be used for evaluation from two angles. The first is how argumentation-based rationale can be used for decision-evaluation by evaluating the consistency and completeness of the rationale as well as evaluating support for development alternatives taking into account decision criteria, input from multiple developers, and uncertainty. The second approach to evaluation describes scenario-based evaluation as supported by scenario-based design (Carroll and Rosson 1992).

Chapter 8, "Support for Collaboration" discusses rationale and collaboration from two perspectives. The first is how the highly collaborative nature of software development supports the development, codification, and use of rationale. The need for collaborators to justify their decisions to each other is a key source of rationale. The other is how rationale supports collaboration by encouraging the exchange of information and awareness of the goals of team members.

Chapter 9, "Change Analysis" identifies the important role that rationale can play in assessing the impact of changing requirements, design criteria, and assumptions on a software system. By explicitly recording the impact that those elements had on the decisions involved and relating the results of the decision-making process to the artifacts that instantiate them, the rationale can be used to detect where changes will be required if requirements, criteria, and assumptions change. In addition, rationale can also capture crucial inter-decision dependencies and alert the developer if one of those dependencies is later violated.

Part 3: Rationale and Software Engineering

In software engineering, decision-making is not restricted to only part of the process. There are critical decisions to be deliberated all throughout the lifecycle of the software system. Part 3 describes how rationale supports the various stages of the software lifecycle and how rationale research relates to other software engineering research that also supports those stages.

Chapter 10, "Rationale and the Software Life-Cycle" gives a brief introduction to the stages of software development and how rationale can be utilized. The topic of life-cycle modeling is then introduced and the application of rationale to sequential models, such as waterfall and the v-model is described as well as how rationale can be applied to iterative approaches. The chapter concludes with a discussion of how rationale supports process improvement initiatives.

Chapter 11, "Rationale and Requirements Engineering" describes rationale's contribution to requirements engineering. This includes how rationale can support the requirements definition process by assisting with requirements elicitation, achieving consensus on requirements, identifying requirements inconsistency, and supporting requirement prioritization. Rationale's role in requirements traceability and the relationship between rationale and non-functional requirements is also described. The chapter concludes with how rationale can assist in adapting to changing requirements, one of the major challenges in software engineering.

Chapter 12, "Rationale and Software Design" describes design rationale as applied to software design. The chapter begins with a description of the nature and importance of software design rationale, both that generated by the designers while designing and that generated during construction and use. Two fundamentally different types of decisions are described—design space decisions and rationale for non-design space decisions that represents a deeper reflection on the design process. We conclude with a look at some specific approaches to rationale as applied to software design and software architecture.

Chapter 13, "Rationale and Software Testing" defines verification and validation and then describes the issues involved in the major types of software tests—inspection, unit testing, integration testing, and system testing. The role of rationale in software testing is described by focusing on three major uses: the contribution of rationale to testability, rationale's contribution to test case prioritization, and using rationale to support component testing and selection.

Chapter 14, "Rationale and Software Maintenance" describes how rationale can be used to support software maintenance. The chapter describes four areas where rationale can support maintenance: maintenance prediction, impact assessment, program comprehension, and maintenance rationale. The chapter then concludes with a discussion of why rationale should also be captured during software maintenance and some existing research that supports the capture of maintenance rationale.

Chapter 15, "Rationale and Software Re-use" begins with a description of key software reuse concepts and categories, along with defining types of rationale that support re-use. The chapter then describes several ways that rationale has been, or can be, applied to assist with software re-use.

Part 4: Frameworks for Using Rationale in Software Engineering

In this section, we take a look ahead. In order to support Rationale-Based Software Engineering, it is necessary to have frameworks to define the key concepts and architectural needs for Rationale Management Systems. In this part, we define a conceptual framework and architectural framework to support Rationale-Based Software Engineering.

Chapter 16, "A Conceptual Framework for Rationale-Based Software Engineering" describes the goals of conceptual frameworks in general, followed by what is needed by a conceptual framework for rationale use in software engineering. To support the decision-centric approaches, we define a taxonomy of software decisions that could be answered using SER.

To support usage-centric approaches, we describe how Carroll and Roson's (1992) Scenario Claims Analysis (SCA) rationale can be applied to software engineering. We conclude with a discussion of the implications of iteration a summary of current challenges to rationale use and propose some potential solutions.

Chapter 17, "An Architectural Framework for Rationale-Based Software Engineering" describes the key features needed for a Rationale Management System (RMS) to support software engineering. This includes the model management subsystem (which includes support for capture and formalization), the underlying hypermedia substrate, and the necessary integrations between RMS and external software development support systems.

Chapter 18, "Rationale Based Software Engineering: Summary and Prospect" serves two purposes. First, it summarizes the work presented in this book and its implications on future rationale research and use. We then look at some key future challenges to software development and conclude with a discussion of both the promises of and challenges to Rationale-Based Software Engineering.

Acknowledgements

This book would not have been possible without the support of many people. First of all we would like to thank to Ralf Gerstner of Springer Germany for making this project possible and for invaluable advice in publishing matters. We appreciate his infinite patience with watching this project come into fruition. We also would like to thank Bashar Nuseibeh and Colin Potts for their excellent and inspiring forewords. At Miami University, Monica Baxter provided secretarial assistance in pulling together the various components of the book. Last, but not least, we would like to thank our friends, family, and colleagues for their support, patience, and encouragement throughout this project.

Author Biographies

Janet E. Burge is an Assistant Professor at Miami University Computer Science and Systems Analysis Department. Dr. Burge's major research interests are in Software Engineering and Artificial Intelligence. Her primary research area is in Design Rationale, with a focus on Design Rationale for Software Maintenance. Prior to her appointment at Miami University in 2005, she taught Software Engineering and Assembly Language at Worcester Polytechnic Institute (WPI) for four years. During and prior to that time, she worked for eight years at Charles River Analytics Inc. on various projects using genetic algorithms for decision-support and on a knowledge elicitation workstation. Before joining Charles River Analytics, she worked for one year at Fidelity Investments developing an expert system to monitor their midrange computer systems and for eleven years at Raytheon Corporation as a software engineer. She received her Ph.D. in Computer Science from WPI in 2005, her M.S. in Computer Science from WPI in 1999, and her B.S. in Computer Science from Michigan Technological University in 1984.

John M. Carroll is the Edward M. Frymoyer Chair Professor of Information Sciences and Technology at the Pennsylvania State University. His research interests include methods and theory in human-computer interaction, particularly as applied to networking tools for collaborative learning and problem solving, and the design of interactive information systems. His books include Making Use (MIT Press, 2000), HCI in the New Millennium (Addison-Wesley, 2001), Usability Engineering (Morgan-Kaufmann, 2002, with M.B. Rosson) and HCI Models, Theories, and Frameworks (Morgan-Kaufmann, 2003). He serves on several editorial boards for journals, handbooks, and series and is Editor-in-Chief of the ACM Transactions on Computer-Human Interactions. He received the Rigo Award and the CHI Lifetime Achievement Award from ACM, the Silver Core Award from IFIP, and the Alfred N. Goldsmith Award from IEEE. He is a fellow of the ACM, IEEE, and HFES.

Raymond McCall is an Associate Professor in the Department of Planning and Design at the University of Colorado, Denver. His major areas of research are in design rationale methods and systems. Since 1992, most of his research has concentrated on the use of rationale to support the design of artifacts for human exploration of space. For much of this time he collaborated with NASA contractors and with employees of the Johnson Space Center in Houston. He has nearly 30 years of experience in design rationale usage in architectural design, planning, policy making and software design. He created the first hypertext systems for support of design rationale in the 1970s and 1980s and was the first to integrate support for rationale capture and delivery into 3D computer-aided design systems. Before coming to the University of Colorado, he worked for six years at the Gesellschaft fuer Information und Dokumentation in Heidelberg, Germany. He received a Ph.D. in Architecture in 1978 from the University of California, Berkeley, and an M.S. in Product Design in 1975 from the Institute of Design at the Illinois Institute of Technology.

Ivan Mistrík is an independent consultant for Software-Intensive Systems Engineering. He is a computer scientist who is interested in software engineering and software architecture, in particular: relating software requirements and architectures, knowledge management in software development, rationale-based software engineering, and collaborative software engineering. He has more than forty years experience in the field of computer systems engineering, primarily working at renowned R&D institutions as a principal scientist and project manager; in parallel he has done consulting on a variety of large international IT-projects sponsored by DFG, ESA, EU, NASA, NATO, and UN. He has also taught university-level computer sciences courses in software engineering, software architecture, distributed information systems, and human-computer interaction. He is the author or co-author of more than 80 articles and papers in international journals, conferences, books and workshops and was the editor of the book "Rationale Management in Software Engineering" published by Springer-Verlag in 2006. In addition, he was the editor of the Special Issue on "Relating Software Requirements and Architectures" published by IEE Proceedings Software in 2005.

Bashar Nuseibeh is Professor and Director of Research in Computing at The Open University (OU), and a Visiting Professor at Imperial College London and the National Institute of Informatics, Japan. Previously he was a Reader at Imperial College London and Head of its Software Engineering Laboratory. His research interests are in software requirements engineering and design, software process modeling and technology, and technology transfer. He has published over 100 refereed papers and consulted widely with industry, working with organizations such as the UK

National Air Traffic Services (NATS), Texas Instruments, Praxis Critical Systems, Philips Research Labs, and NASA. He has also served as Principal Investigator on a number of research projects on software engineering, security engineering, and learning technologies. Bashar is Editor-in-Chief of the Automated Software Engineering Journal, and an Associate Editor of IEEE Transactions on Software Engineering, the Requirements Engineering Journal, and a number of other journals. He was a founder and first Chairman of the BCS Requirements Engineering Specialist Group (1994-2004), and is currently Chair of IFIP Working Group 2.9 (Software Requirements Engineering) and Chair of the Steering Committee of the International Conference on Software Engineering (ICSE). He has served as Programme Chair of major conferences in his field, including the ASE'98, RE'01, and ICSE-2005. Bashar holds an MSc and PhD in Software Engineering from Imperial College London, and a First Class Honours BSc in Computer Systems Engineering from the University of Sussex, UK. He received a Philip Leverhulme Prize (2002), an ICSE "Most Influential Paper" award (2003), a "Best Application Paper" award from the 18th International Conference on Logic Programming (ICLP-2002), and a number of other best paper and service awards. He held a Senior Research Fellowship of the Royal Academy of Engineering and The Leverhulme Trust between 2005-2007. He is a Fellow of the British Computer Society (FBCS)and the Institution of Engineering and Technology (FIET), an Automated Software Engineering Fellow, and is a Chartered Engineer (C.Eng.).

Colin Potts is Associate Professor in the School of Interactive Computing at the Georgia Institute of Technology, where he conducts research in human-centred computing; specifically in requirements elicitation and analysis, feature clustering and evolution, privacy-aware computing, and design rationale. He is also active in the teaching of social implications of computing and professional ethics. Before joining Georgia Tech, Colin Potts worked at Imperial College, London and the Microelectronics and Computer Technology Corporation. He has a PhD in psychology from Sheffield University.

Contents

Part 1
Introduction

Rationale research, which has been going on since the 1970s, initially focused on Design Rationale – the reasons behind decisions made when designing. This is an appropriate term in many domains where a physical artifact is first designed and then manufactured. While there is a phase in most SE lifecycles that produces a software design (design as a noun), the act of designing (design as a verb)—making the decisions that affect that design and how it is realized in the software system—takes place throughout the software development process. In order to make this distinction clear, in this book we refer to rationale as Software Engineering Rationale (SER), as defined in Dutoit et al. (2006) and refer its use as a key aspect of the software process as Rationale-Based Software Engineering (RBSE).

The first step towards RBSE is an understanding of what rationale is and how it can help us meet the critical challenges that software engineering faces (Chapter 1). Software is not the same as hardware and these differences affect both what the rationale is (structure and content) and how rationale can be used (Chapter 2). These differences provide both opportunities, such as the ability to directly link rationale to the artifacts that it describes, and challenges, such as the need to support iteration.

SER can have many roles in supporting software engineering (Chapter 3). The decisions where rationale should be captured include not only those occurring during development but also those affecting the choice of software development process/methodology, management strategy, and how the software will be verified, validated, and even deployed.

The rationale research described here builds on work that started with Rittel's Issue-Based Information system (IBIS) (Kunz and Rittel 1970), initially applied to urban planning. Those proposing approaches to applying rationale to software engineering would be doing their research a disservice by not learning from the experience of applying rationale to other domains (Chapter 4). And finally, it is important to understand that decision-making, in particular human decision-making lies at the heart of software engineering and how Rationale-Based Software Engineering supports that process (Chapter 5).

1 What is Rationale and Why Does It Matter?

As the term is used here, rationale is the reasoning underlying the creation and use of artifacts. Software engineering research on rationale aims to devise methods and systems for managing rationale throughout the software engineering process. Managing rationale includes eliciting it, recording it, indexing it for retrieval, editing it, and retrieving it for those who need it. Recorded rationale can play a valuable role in every stage of the software lifecycle and for every participant in that lifecycle. It can help developers to create better software by enabling them to learn from the successes and failures of the past. It can facilitate coordination and collaboration within development teams, aid in identification and analysis of requirements, as well as design, testing and maintenance. It can even help users to understand the systems they use.

1.1 Introduction

1.1.1 The Scope and Value of Rationale in Software Engineering

As used here, the term *rationale* denotes the reasoning underlying the creation and use of artifacts. Rationale research seeks ways of aiding decision makers by creating explicit records of this reasoning. Most other types of research on decision making, by contrast, seek to create formal, computational methods for deriving decisions. Rationale research *primarily* deals with informal and semi-formal, verbal reasoning; but it does not ignore formal reasoning and computation, both because humans sometimes use these in reasoning about decisions and because they can augment human reasoning. While rationale is primarily verbal, various kinds of graphics can play crucial supporting roles. Not all rationale can be made explicit. Nevertheless, researchers generally appear to believe even incomplete records of rationale can improve the quality of artifacts.

To date, almost all research on rationale in various application domains has dealt solely with design rationale, i.e. the reasoning within the design process. In fact, the term *design rationale* is often used as if it were the only subject of rationale research. But to understand the full meaning and importance of the term *rationale*, one must look further. Design is only part of the larger process of artifact creation, and rationale-based decision making is found in every other part of that process. It is found, for example, in the determination of requirements, the construction of the artifact, the maintenance of that artifact and the administration of the overall creation process.

This chapter and this book deal with the full potential of rationale in software engineering (SE), i.e. not only in design but in all parts of the software lifecycle and all aspects of SE. Since the term *design rationale* does not encompass the full scope of reasoning about decisions in SE, the term *software engineering rationale (SER)* is used here instead. *Rationale-based Software Engineering (RBSE)* research investigates concepts, theories, approaches, methods and software needed to realize the full potential of SER to aid SE. The typical approach to realizing this potential is to create *rationale management systems* (RMSs), i.e. software that aids in the elicitation, recording, structuring, indexing, retrieval and distribution of SER to stakeholders in software projects.

This chapter argues that a rationale-based approach will be essential for meeting the current and future challenges of SE. Software developers and maintainers currently find themselves deluged with problems that severely tax their abilities, and yet the future looks even more challenging. Many current software projects fail completely, and many others achieve only partial success. Nevertheless, software projects continue to grow relentlessly in number, variety, scale, complexity, longevity and technological requirements as developers attempt to keep up with competitors, customer demands and new hardware capabilities.

Software engineers are currently wrestling with the issue of how SE will need to adapt to meet the challenges of the future. We argue that these challenges make it crucially important that participants in software projects understand the reasoning of others involved in such projects. Absence of such understanding creates the risk of serious errors in requirements, design, implementation, maintenance, redesign, coordination and project management. Achieving such understanding requires the use of software engineering tools that manage rationale.

1.1.2 Objectives of this Chapter

The first objective of this chapter is to explain what rationale is. The second is to explain how SE will derive crucial benefits from a rationale-based approach. To explain what rationale is, Section 1.2 provides a rough sketch of research in rationale. To explain why rationale matters for SE, Section 1.3 begins by looking at ways in which rationale can be useful for artifact creation in general. It then lists various ways in which rationale can aid software engineering in particular. It looks at the problems facing future software engineers and discusses ways in which rationale management can alleviate these problems. Finally, Section 1.4 summarizes the chapter and indicates where to find further information on the nature of rationale research in SE and why it matters.

1.2 A Rough Sketch of Research on Rationale

The general goal of rationale research is to use records of rationale to improve the processes of creating artifacts of various kinds, including physical artifacts such as buildings, cities and machines as well as cognitive artifacts such as software and governmental policy. To support this goal rationale research has sought to develop methods and software that enable

- the elicitation of useful rationale from its authors
- the recording of useful rationale
- the structuring and indexing of rationale to aid its retrieval
- retrieval of rationale when it is useful
- delivery of that rationale to those for whom it is useful and
- use of the rationale by those people

A good way to get a rough preliminary understanding of subsequent approaches to rationale management methods is to view them as either *variations* on Issue-Based Information Systems (IBIS) (Kunz and Rittel 1970) or *fundamental alternatives* to it. This implies comparing these approaches to IBIS. This, of course, pre-requires at least a basic understanding of IBIS—which is where we shall begin.

1.2.1 Argumentative Approaches to Rationale

Rittel's pioneering work on design rationale was motivated by his theory of *wicked problems* (Rittel and Weber 1972), an idea that has also influenced

SE in other ways (Budgen 2003; De Grace and Stahl 1998; Fitzpatrick 2003). Rittel saw design problems as *wicked* in the sense that they presented fundamental difficulties that could not be overcome using either strictly scientific methods or purely automated methods such as those of artificial intelligence and optimization theory (Rittel 1972). Instead, they required new types of methods that supported creative human problem solving (Rittel 1980) by means of what he called an *argumentative approach* (Rittel 1972). In this approach, every step in problem solving can be seen as part of an inquiry that involves questioning, proposing ideas and subjecting them to critical discussion from the viewpoints of different stakeholders. Rittel devised IBIS to implement this argumentative approach. A number of other rationale methods have either modified Rittel's approach or invented their own argumentative methods from scratch. Procedural Hierarchy of Issues (PHI) (McCall 1979; McCall 1991) and Decision Representation Language (DRL) (Lee 1991) and RATSpeak (Burge and Brown 2006) are examples of the former. Questions Options and Criteria (QOC) (MacLean et al. 1990) and Scenario-Claims Analysis (SCA) (Carroll and Rosson 1996) are examples of the latter.

1.2.1.1 IBIS

IBIS structures rationale using a fixed conceptual schema featuring given element types and given relationships between them. The schema divides rationale into processes of deciding various *issues*, stated in the form of questions. Proposed decision alternatives for an issue are called *positions*, and reasoning about the merits of the positions is represented as *arguments* for or against 1) the positions or 2) other arguments. The decision taken on an issue is its *resolution*. Relationships of various kinds link issues to each other. Figure 3.1 shows how IBIS could be used to document preliminary discussion on one issue in a project on creation of a Rich Internet Application. In addition to dealing with design of system features, *issues* in IBIS can deal with any other topic in artifact development that can be phrased as a question to be answered.

IBIS has most often been used to structure design discussion as it takes place. But at times it has been used to retrospectively give structure to free-form design discourse. Sometimes these retrospective descriptions reflect the actual history (temporal sequence) of the discussion; sometimes they are idealized accounts that ignore history in favor of a more "logical" organization of the rationale. The former is called a *process-oriented approach*; the latter a *structure-oriented approach*.

Certain general features of IBIS are shared by most other argumentative approaches to rationale. These include the following:

Issue: <u>What programming technology should we use to implement</u> <u>the client for our Rich Internet Application?</u>

    Position 1: AJAX (Asynchronous JavaScript and XML)

        Arguments on this position:

            Against: AJAX still has problems with some older browser versions.

            For: This approach makes good use of W3C standards.

                Arguments on this argument:

                    Against: AJAX makes it difficult to meet the guidelines of the W3C's Web Accessibility Initiative.

                    Against: Some kinds of AJAX use in-line frames, which are not part of the W3C's XHTML 1.1 recommendation.

    Position 2: Flex/Flash (with ActionScript)

        Arguments on this position:

            For: Flash has more than 98% browser penetration.

            For: Flash is almost completely platform independent.

            Against: Flash technology is proprietary and thus could change rapidly in ways that would be detrimental to our project. Public standards tend to put the brakes on such rapid change.

                Arguments on this argument:

                    Against: Flash's enormous installed base makes it extremely unlikely that it would change in such a way as to break existing applications.

                    Against: The ActionScript Virtual Machine was donated to the Mozilla Foundation and is now the basis for the Tamarin open source project.

                    Against: Flex is also an open source project.

    Position 3: Silverlight

        Arguments on this position:

            For: Silverlight will work across IE, Firefox and Safari browsers.

            For: Silverlight is compatible with AJAX and can make AJAX development easier through use of the Atlas technology.

Fig. 3.1. Partial discussion of one IBIS issue is shown here in outline format.

1. Using a fixed conceptual schema of elements and relationships between pairs of them
2. Dividing rationale into the reasoning about individual decision-making tasks (called *issues* in IBIS)
3. Representing decision-making tasks as questions to be answered
4. Proposing decision alternatives for each decision-making task (called *positions* in IBIS)
5. Evaluating the proposed decision alternatives by stating and considering pros and cons of these alternatives (called *arguments on positions* in IBIS)
6. Evaluating the evaluations by stating and considering pros and cons (called *arguments on arguments* in IBIS)
7. Deciding a decision task by selecting one decision alternative on the basis of its evaluation
8. Using several relationships to link the separate decision-making processes (called *inter-issue relationships* in IBIS)

We will look briefly at a number of other argumentative methods:

- PHI (Procedural Hierarchy of Issues) (McCall 1979) (McCall 1991),
- revisions of IBIS by Potts and Bruns (Potts and Bruns 1988) (Potts 1996)
- QOC (Questions, Options & Criteria) (MacLean et al. 1996)
- DRL (Decision Representation Language) (Lee 1991) (Lee and Lai 1996)
- RATspeak (Burge and Brown 2006)
- Scenario-Claims Analysis (Carroll and Rosson 1996) (Carroll 2000)

All except the last of these approaches can usefully be viewed as variations on some of the ideas introduced by IBIS, though it should be noted that QOC was not derived from IBIS. Scenario-Claims Analysis represents a fundamental departure from the other approaches.

1.2.1.2 PHI

PHI is a refinement of IBIS and its main innovation is to show that frequently the decision on one issue depends on the decisions made on others. For example, the decision on the issue in Figure 3.1 could depend on the decision on the issue, "Is it important that our project adhere to

W3C standards?" PHI models rationale as a connected graph of issues linked by such dependency relationships. This structure tends to be roughly hierarchical, thus the name Procedural Hierarchy of Issues, and has a *root issue* representing the project as a whole. The root issue of the project for Figure 3.1 might be, "What is our web-based CAD system to be?" Such an issue has three crucial properties: 1) the process of deciding this issue *is* the development project in its entirety, 2) the final decision on this issue is a representation of the final, constructed artifact, and 3) the decision on this issue depends on the decisions to all the other issues in the project.

1.2.1.3 Potts and Bruns

Inspired in part by Conklin's and Begeman's use of IBIS (Conklin and Begeman 1988) in their gIBIS hypertext system, Potts and Bruns (1988) modified IBIS for use in software design. The crucial innovation of their approach was to include in their schema elements that represented "intermediate artifacts," i.e. the various models and documents produced during design to represent the software being designed. In other words, their schema was not exclusively a *rationale* schema, but rather a hybrid schema containing both rationale and artifact elements. This approach created design histories in the form of a collection of linked intermediate-artifact and rationale nodes.

Besides adding intermediate-artifact nodes, Potts and Bruns also made some modifications to IBIS itself. Instead of having separate *argument* elements, they represented all argumentation on a given decision (issue) in a single *justification* statement. They also argued that to put IBIS to practical use in software design, it would have to be tailored to specific software design methods. They give an example of this that shows how it could be adapted to work with Liskov and Guttag's software design method (1986).

Potts (1996) went on to elaborate the original Potts and Bruns approach to give a general account of how IBIS-based rationale could be used to support software methods. In particular, he argued that methods could themselves be modeled as recurring, method-specific collections of issues combined with method-specific types of intermediate-artifact nodes. He supported this argument by providing accounts of how three specific software methods could be represented in this manner.

The Potts and Bruns approach was to inspire the creation of DRL (Lee 1990), which in turn inspired the creation of RATspeak (Burge and Brown 2006). Many SE-specific approaches to rationale also adopt the idea originated by Potts and Bruns of using hybrid artifact-rationale schemas (See Chapter 12: Rationale and Software Design), and a number of the recommendations of this book center on this idea as well (See, for example, Chapter 17: An Architectural Framework).

1.2.1.4 QOC

Like IBIS, QOC centers on decision tasks that are represented as *questions* and evaluates proposed decision alternatives, called *options*. QOC, however, only deals with "design space" questions, i.e. those that determine features of the designed artifact. Thus, there are many issues that IBIS and PHI can deal with but that QOC does not attempt to deal with.

QOC's main innovation is a finer level of granularity of elements in the evaluation of alternative answers to questions. Instead of an IBIS-type *argument* on a proposed alternative, QOC uses a pairing of a *criterion* and an *evaluation of the alternative with respect to the criterion*. This is especially significant for software engineering, because 1) software requirements can be represented as criteria and 2) doing so *enables the tracing of requirements to specific features of the artifact*. Also significant is the fact that two other rationale methods discussed in this section, namely DRL and Scenario-Claims Analysis, have also opted for the QOC style of evaluation rather than the IBIS style. QOC and DRL do, however, allow IBIS-type arguments for and against the criterion-based evaluations.

The authors of QOC do not use the method to model rationale as it is being generated and do not attempt to structure designers' thinking processes using the QOC schema. Instead, they use the method merely for retrospective documentation of design rationale. In other words, they use a structure-oriented rather than a process-oriented approach.

1.2.1.5 DRL

DRL revises and extends the approach of Potts and Bruns (1988). DRL's schema corresponds roughly to a superset of QOC's that also has dependency relationships between elements, including some derived from PHI. DRL has a finer-grained schema than other approaches and is thus more "expressive." While it does not always represent decision tasks as questions, DRL uses a semantically equivalent form. Examples of DRL in the literature deal only with the "design space" decisions like QOC, but it seems in principle that DRL could be applied to the larger range of decision tasks dealt with by IBIS.

In devising DRL Lee objected to the Potts and Bruns approach of merging the many arguments of IBIS into a single "justification" statement. But Lee also abandoned the *argument* category in favor of one called *claims*.

At first, a *claim* appears to be a single sentence rather than the multi-sentence, multi-premise syllogism that the laws of logic require an *argument* to be. This might give the erroneous impression that arguments are being broken into their constituent premises (*claims*). In reality, however, the DRL claims that are used to evaluate other claims are *enthememes*, that is, multi-premised arguments in which all but one of the premises are left unstated because their existence is clear from common sense. In other words, such *claims* are really just an elliptical form of argument in which only one of the premises is stated. In fact, enthememes are common in all argumentative discussion, regardless of which rationale schema is used. The problem with replacing the *argument* category with the *claim* category is that it is not at all clear that all arguments can be stated as enthymemes.

1.2.1.6 RATSpeak

Burge and Brown (2006) describe RATSpeak as an extension of DRL that is designed to make it more suitable for use in software engineering. But some of the RATSpeak revisions make it more like IBIS. In particular, it reinstates *arguments* as elements of the rationale schema in addition to *claims*.

RATSpeak introduces a number of new categories of elements into its schema to enable a greater amount of automated checking and inference-making than would be possible with DRL. For example, *requirements, assumptions* and *background knowledge* are introduced as element types. In addition, it adds an *argument ontology* consisting of a hierarchy of argument types tailored to the domain of software engineering. This ontology is used for automated checking of the rationale for correct form.

RATSpeak contains a crucial innovation in the form of a special type of argument that describes dependencies between alternatives on different decision tasks. These arguments enable the description of how adoption of an alternative on one decision might help or hinder the adoption of an alternative on a different decision. No other argumentative approach described here enables the recording of such dependencies between alternatives—though PHI and DRL can represent dependencies between decision tasks. This makes RATSpeak especially valuable for change analysis and iterative approaches to software engineering. It also suggests that other argumentative schemas are inherently incomplete.

1.2.1.7 Scenario-Claims Analysis

Scenario-Claims Analysis (SCA) is strictly for design of human-computer interaction. It uses scenarios of software use to evaluate system features with respect to users' goals. Like QOC and DRL, SCA evaluates features as positive or negative with respect to goals, but unlike other approaches, *it does not represent decision tasks or decision alternatives*. It only represents 1) system features, 2) use-based criteria/goals by which they are evaluated and 3) users' positive or negative evaluations of the features against those criteria. It does not use deeper levels of argumentation on these evaluations; but since these goal-based evaluations are themselves arguments for or against the features, SCA must be counted as an argumentative approach. SCA is the only rationale method to deal explicitly with usage scenarios; as such, it is the most user-centered of the rationale approaches. It is also the only argumentative approach to show how rationale fits in an iterative process of design. In short, SCA represents a fundamentally different view of rationale for design.

1.2.2 Rationale Methods that Go Beyond Argumentation

1.2.2.1 Structuring Rationale Using Artifact Structure

One common way of documenting rationale uses the structure of the designed artifact rather than the structure of an argumentative schema to organize rationale. In fact, this is simplest and least labor-intensive way of recording rationale (Lee and Lai 1996). In design of physical artifacts, this can be done by simply linking textual rationale to a digital model of the artifact being designed, as has been done by Reeves (Reeves and Shipman 1992) and by Domeshek and Kolodner (1996). In software development, this can be done by linking textual rationale directly to sections of code as has been done by Schneider (2006). It should be noted that some schema-based software systems for argumentative rationale, such as SEURAT (Software Engineering Using RATionale) (Burge and Brown 2006), also make it possible to link rationale directly to the artifact, which in the case of SEURAT is the source code. This shows that there is no inherent conflict in structuring an artifact both around its argumentative structure and the structure of the artifact.

1.2.2.2 Problem-Based Evaluation

Lewis, Riemann and Bell (1996) present a novel approach for evaluating alternative features of an artifact. They describe their own software design process as using *a suite of problems* for conceptual testing of different proposals for a programming environment they were designing. They evaluate a design proposal by looking at how well it could be used to solve the problems in the suite. Their work suggests that argumentation may not be the only, or even the best, means of evaluating alternatives in all cases. This challenges the sufficiency of the argumentative approaches that currently dominate rationale research. While the notion of problem-based evaluation suggests an interesting direction for future work on rationale in software design, it does not yet constitute a generally applicable rationale method.

1.2.2.3 Generative Rationale

Gruber and Russell (1996) argue that argumentative schemas prejudge what information will be needed later by software engineers. No advance collection of rationale, they claim, could answer all of the questions that might later be raised about the rationale for an artifact. Rather than having designers create highly detailed models of their rationale, they argue that it would be better to collect engineering data and models during the project and then later use these to deduce the rationale behind the artifact in response to questions that arise about it. This claim constitutes an interesting hypothesis about possible future approaches to rationale for software development; but, as is the case with the above-described problem-based evaluation, it does not yet constitute a generally applicable rationale method.

In the end, then, it is primarily to the argumentative approaches that we must look for viable approaches to documenting design rationale. The one non-argumentative approach that is viable is that in which rationale is structured according to the structure of the artifact being designed, e.g. the software. Typically, this means that pieces of the rationale that discuss a part of the artifact are associated with that part of the artifact in some way. There are several ways in which this can be done. One is by linking parts of the rationale to corresponding parts in a model of the artifact. Another way to do this in software development is to link parts of the rationale to parts of the artifact itself, i.e. either in the source code or the run-time (compiled/interpreted) code.

1.3 Why Rationale Matters

Rationale matters because it is useful for artifact creation in general and for software engineering in particular. We will first look at the former and then at the latter.

1.3.1 The Usefulness of Rationale for Artifact Creation

There are two ways in which rationale documentation methods can be useful for artifact creation. The first and most basic is by providing a record of the reasoning associated with decision making. The second is by actively shaping the process of reasoning about decisions. The following sections look at each of these in turn.

1.3.1.1 The Usefulness of Rationale as a Record of Decision Making

There are two respects in which a record of the reasoning in decision making can be useful. One is by serving as a memory aid for those who participated in the decision making. The other is by informing those who did not participate but are affected by the decisions.

Rationale as memory aid. For those who participate in making given decisions, having records is important because of a tendency to forget what was decided and why. Correctly remembering all the rationale for the decisions in a project is generally more than any individual can do. This is especially true in large and complex projects, in which hundreds or even thousands of decisions are made.

Documented rationale also provides a crucial resource in case decisions need to be revisited. This often happens when, after decisions have already been made, they are discovered to have undesirable consequences. It can also happen when new features need to be added to an artifact, or when the artifact needs to be redesigned. On the other hand, documented rationale can prevent decisions from being pointlessly being revisited, which can happen when the rationale for them is forgotten.

Sometimes artifact creators find themselves facing decision tasks similar to ones in prior projects they have worked on. In such cases, they often feel that it would be useful to know how they arrived at those previous decisions. Unfortunately, they also often find that they cannot precisely recall their own prior rationale. Documented rationale serves as a valuable memory aid in these cases as well.

Rationale as information for other stakeholders. Documenting rationale for decisions can also help people who do not participate in making those decisions but who nevertheless have a stake in what is decided. Such stakeholders include those who must implement the decisions, those who need to coordinate their own decision making with the given decisions and those who manage the processes of artifact development and maintenance. The rationale for a given decision typically indicates what the goals and evaluation criteria of the decision makers are; this information enables others participating in development or maintenance to make sure their own efforts do not conflict with those goals and criteria.

One important use of rationale as information exists when people join or leave a decision-making team. There is a tendency for new team members to challenge decisions made before they arrived. This can be unnecessarily disruptive if the new team members base their challenges on ignorance of the rationale for decisions. It is therefore useful to require new team members to examine the documented rationale for decisions before challenging them. And when people have left a project, it is no longer possible to ask them about their rationale for past decisions; so having documentation of their rationale becomes crucial.

1.3.1.2 The Usefulness of Rationale as an Aid to Decision Making

In addition to the value of simply recording it, rationale can be useful by aiding decision making. There are two, mutually compatible approaches to doing this. One is by providing information that helps people to make better decisions. Since this information aids decision making, it is by definition rationale. The goal of this first approach might be described as *informed decision making*. The information used to aid design might come from feedforward from earlier decisions or feedback from later decisions, implementation or use of the artifact. It might also come from previous projects.

The second approach to aiding decision making is to prescribe the process by which the reasoning about decisions proceeds. This is typically done in an effort to make rationale more thorough, consistent, or carefully argued. The goal of this approach might be described as *well-reasoned decision making*. Typical procedural prescriptions include making sure that alternatives are considered for every issue, that all such alternatives are evaluated by the relevant criteria, that the arguments both for and against the alternatives are considered and that the argumentation is representative of the full range of stakeholders.

1.3.1.3 The Descriptive and Prescriptive Roles of Rationale

With respect to *any given decision or set of decisions* a rationale approach can play a purely descriptive role, a purely prescriptive role or a combined prescriptive-descriptive role. In a purely descriptive role, a rationale approach merely seeks to record the reasoning of decision makers and does not seek to influence that reasoning or the decisions made. When used in this way with respect to some decision making tasks a rationale approach typically seeks to improve other types of decision making tasks. For example, it is common to find approaches that seek to record the rationale of designers not to influence the design but to influence construction, maintenance or project management.

When a rationale approach seeks to influence a given type of activity, such as requirements determination or project management, then it is, by definition, prescriptive with respect to that activity. As a consequence, rationale approaches are generally prescriptive with respect to one or more such activities.

A rationale approach would be *purely prescriptive* with respect to a given set of decisions if it only sought to improve the reasoning of the associated decision makers without keeping records of their reasoning. For example, QOC records design rationale as a way improving the decision making in software construction, yet it does not record the reasoning of software architects or programmers who do the decision making about construction. QOC is thus purely prescriptive with respect to the decision making in construction. Similarly, Scenario-Claims Analysis uses the rationale of users to inform the design of human-computer interaction (HCI) but does not record the reasoning of those who design this interaction. It thus plays a purely prescriptive role with respect to HCI decision making.

A rationale approach is both prescriptive and descriptive with respect to decisions when it seeks both to influence the reasoning of the associated decision makers and to record their reasoning. IBIS, for example, is often used as a procedure for running design meetings to make sure decision makers look at a wider range of decision alternatives (positions) and wider range of arguments both for and against the proposed alternatives. And having elicited this wider range of rationale, IBIS makes sure that it is documented so that it is not lost. When used in this manner IBIS is prescriptive-descriptive with respect to design decisions.

It should also be pointed out that a rationale approach might start out merely as a way of describing the rationale for a given set of decisions— such as design decisions—then later become prescriptive with respect to that same set of decisions by serving as a memory aid for the decision makers—i.e. the designers.

1.3.2 The Usefulness of Rationale for Software Engineering

1.3.2.1 Possible Uses of Rationale in Software Engineering

One way to quickly get an idea of the value of rationale for software engineering is to look at the range of its possible uses. The outline shown below is by no means complete, but it gives an idea of this range:

Uses of Rationale in Software Engineering:
- **Supporting requirements engineering**
  1. Supporting identification of requirements
  2. Supporting explanation/evaluation of requirements
  3. Supporting revision of existing requirements
     a. By providing feedback from design, implementation and use
  4. Supporting addition of new requirements
     a. By providing feedback from design, implementation and use
  5. Helping requirements engineers make better decisions by informing those decisions and improving the reasoning underlying them
     a. By providing records of the decisions and reasoning about requirements from past projects
- **Supporting design**
  1. Providing traceability of requirements to design decisions and vice versa
  2. Helping designers make better decisions by informing those decisions and improving the reasoning underlying them
     a. By providing feedback from implementation, maintenance and use to validate requirements
     b. By providing rationale behind design patterns
     c. By providing records of the design decisions and reasoning from past projects
- **Supporting implementation of software**
  1. Providing traceability of requirements and design decisions to implementation decisions and vice versa
  2. Helping implementers make better decisions by informing those decisions and improving the reasoning underlying them
     a. By providing rationale for implementation patterns
     b. By providing records of the implementation decisions and reasoning from past projects

- **Supporting software maintenance**
  1. Helping maintainers to make better decisions by informing those decisions and improving the reasoning underlying them
     a. By helping maintainers to understand the rationale for the requirements of users, the decisions of designers and the implementation decisions of programmers
     b. By providing feedback from the use of the software to make it clear when maintenance is needed
     c. By providing historical records of the rationale for maintenance decisions
- **Supporting project management**
  1. Making it possible for managers to understand when decisions are being made by various participants in the software project, why those decisions are being made and who is likely to be affected by those decisions
  2. Helping project managers to make better decisions by informing those decisions and improving the reasoning underlying them
     a. By providing historical records of the rationale for management decisions
     b. By providing records of the rationale for management of past projects
- **Supporting use**
  1. Providing rationale as explanations of the functioning of complex software systems
- **Supporting the work of groups**
  1. Using rationale as a vehicle for communication amongst different kinds of experts and stakeholders in a project
  2. Exposing differing points of view amongst stakeholders
  3. Facilitating participation of stakeholders and collaboration among team members by making the decision making process "transparent" and open to inspection
  4. Making it clear when the decisions of a given group of people supports or interferes with the decisions of others
  5. Building consensus
     a. By providing greater transparency—nobody is hiding anything
     b. By exposing conflicting points of view early in the process so that they can be negotiated
     c. By revealing areas of agreement, so that they can serve as starting points for building consensus

- **Supporting change**
  1. Helping to detect when change is needed
     a. By providing a record of assumptions that could become invalid in the future, including assumptions about facts, requirements, means, constraints and evaluation criteria
     b. By providing feedback that shows when decisions have produced unforeseen consequences that in turn suggest revisiting and revising decisions
     c. By providing information from users that indicate new or newly discovered requirements
  2. Helping to cope with current changes and to prepare for future changes
     a. By showing the network of dependencies among decisions that indicate how the effects of a given design change can ripple through the design of the software
     b. By showing which team members' work will be affectted by changes
     c. By showing the goals and evaluation criteria for the current version of the software, and thus indicating goals and criteria that a redesigned system should also satisfy
     d. By providing a record of decision alternatives and their evaluations to facilitate the redesign of the system
  3. Supporting the management of change, by showing its affects on the work of individuals and groups as well as on the expenditure of time and money
- **Supporting software reuse**
  1. Providing explanations for what code is designed to achieve as well as why it is designed and implemented the way it is
- **Supporting knowledge transfer**
  1. Enabling learning from the successes, failures and ideas of past software projects
  2. Validating designs
  3. Collecting, organizing and delivering reusable knowledge for development and maintenance
  4. Supporting training and education
  5. Supporting research on real-world software engineering projects

1.3.2.2 Rationale and the Future of Software Engineering

To understand why rationale matters for SE, it is not enough to know the range of its possible SE applications. It is also crucial to know what the value of these applications are in view of the profound challenges now facing the SE field.

The challenges that software engineering faces. The current state of software development is not good. Developers have been unable to keep up with the dramatic progress in hardware resulting from Moore's Law and the spread of the Internet throughout the world and into every aspect of people's lives. It is not enough to urge software developers to do a better job, because they already find themselves coping with difficulties that tax their abilities severely.

Unfortunately, current trends indicate that the future will be even more challenging. Software progress will increasingly lag behind hardware progress. As developers scramble to keep up with new technologies, rising customer expectations and aggressive competitors, they will find that the development tasks they face are getting progressively more difficult. Software projects will continue to grow relentlessly in number, variety, scale, complexity and longevity. This will make coping with any given amount of change increasingly difficult, but it will also increase dramatically the amount of change that must be coped with.

The issue of the increasing longevity of systems by itself represents in a microcosm the future difficulties awaiting developers. This increasing longevity results from the initial success of software systems. Successful systems stay on the market and go through version after version. This long life of systems creates a host of problems. As systems get older they tend to increase in functionality, driven by the pressure of new hardware capabilities, increases in user expectations for functionality and the need to keep up with competitors. Additional functionality increases the size and complexity of systems, thus making it progressively more difficult to maintain the systems and add new features without breaking existing functionality or angering the installed bases of users who are used to previous versions. Typically, systems grow in this incremental manner until further growth becomes too difficult, at which point the systems are comprehensively redesigned and re-implemented.

The picture that emerges is one in which software goes through many cycles of redesign and re-implementation over the many years of its life. Each such cycle creates the dangers that 1) good ideas in the system's design and implementation will be lost and that 2) hard-won lessons about how *not* to design and implement the system will be forgotten. Predicted future increases in software scale and complexity increase these dangers dramatically.

In addition to the above-listed problems, Patterson (2005) has argued that future creators of software face additional challenges due the legacy of a 20th Century value system that is profoundly unsuited to 21st Century software development. In the 20th Century the priorities were faster and cheaper computers and communication. These priorities generated the current problems of lack of security, privacy and reliability. Of course, *cheaper* software only meant cheaper to purchase, not cheaper to install, operate and maintain. Adding to these problems is the fact that the increased speed and capacity of systems was used almost exclusively to add features to software rather than make it easier to use. Patterson therefore calls for 21st Century developers to abandon their 20th Century values of cost and performance and in favor of what he calls the "SPUR" challenges: security, privacy, usability and reliability.

But the SPUR challenges are not the only problem. Patterson points out that there are many other crucial challenges facing 21st Century developers. He lists two important examples:

- Extending web search to all information, including multimedia information, and to all people, including those outside the first world
- Adapting software, including operating systems, programming languages, databases and applications, to massively parallel microprocessors.

The upshot of the various challenges that Patterson identifies is the necessity for fundamental redesign of almost all the software currently in use, as well as redesign of the software engineering process itself to support the new priorities of 21st Century software.

How rationale can help in meeting these challenges. Software engineering appears to be headed into a future characterized by incessant change and repeated redesign of software systems that have grown greatly in scale and complexity. Change and redesign invariably create the risk of side-effects that damage the quality of a system, as for example, when existing good features become lost or broken. But such unintended and undesirable side-effects can be much more easily be avoided if those making changes to the system *understand the rationale underlying the systems they are changing.* The justifications for existing features help re-designers and maintainers to understand what aspects of a system need to remain constant and how a system can change while still achieving the goals of previous versions.

When development teams are small and when the history of the software is short, there might appear to be little need for documentation of the rationale behind the decisions that went into its creation. Those who change the systems are likely to be the same people who created it to begin with. In such cases, the rationale underlying the system can be accessed through memory and informal communication with other project participants.

But when software is older, larger and more complex, the need for documentation of rationale is more obvious, because it becomes difficult or impossible to know all the system rationale without documentation. The people doing redesign, reimplementation and maintenance are unlikely to be members of the original development team. Without documentation of rationale, the current developers and maintainers have little or no access to the rationale of those who worked on the system previously. Without knowledge of this rationale, the chances are great that redesign and reimplementation will result in serious errors and that attempted improvements will actually degrade the system.

Future software developers and maintainers will be greatly aided in their work if they have an understanding of the rationale behind the systems they seek to improve. Tools that provide this understanding must be integrated into the environments that software engineers use to create and maintain software. In particular, these tools must be capable of managing and delivering relevant rationale to software engineers when and where they need it. A central goal of rationale research in SE is to create tools that enable the use of rationale throughout the SE process and thus make possible a rationale-based approach to software engineering.

1.4 Summary and Conclusions

Rationale research studies the reasoning underlying the creation and use of artifacts. It seeks ways of aiding decision makers by creating, storing and retrieving explicit records of this reasoning. While this research has until recently focused almost exclusively on rationale for design, attention has begun to shift to the many other parts of the artifact lifecycle where rationale-based decision making plays crucial roles. To understand fully what rationale is and why it matters, it is necessary to understand all of these roles.

Starting with Rittel's IBIS (Kunz and Rittel 1970), the dominant theme in rationale research has been modeling the argumentative structure of rationale. Almost all argumentative approaches—including IBIS, PHI, the

Potts-Bruns approach, QOC, DRL and RATspeak—have modeled the evaluation by decision makers of decision alternatives using argumentation. Scenario-Claim Analysis has been unique in modeling the evaluation by users of features of designed artifacts during scenarios of artifact use. More detailed treatments of these and many other approaches are found in other chapters of this book. In addition, a detailed overview of current research on rationale in software engineering can be found in the book, *Rationale Management in Software Engineering* (Dutoit, McCall, Mistrik and Paech eds. 2006).

Rationale matters for SE, first of all, because it has a wide spectrum of uses to aid decision making and other activities throughout the software lifecycle. But in addition, rationale matters because the ways in which it aids SE have great value in meeting the profound challenges that are facing the future of software engineering. In particular, rationale is especially useful for dealing with large-scale, high-functionality software projects characterized by constant change and repeated redesign.

2 What Makes Software Different

Research on rationale in software engineering was originally inspired by research on rationale for the design of physical artifacts. While there is still much that software engineering can learn from the latter, it is important to recognize that the process of software development differs in crucial ways from the processes of developing physical artifacts. These differences have important consequences for the successful implementation of rationale management. One consequence is that software development has unique and urgent problems that rationale management can do much to solve. Another is that the ways in which software differs from a physical artifact provide unique advantages for implementing rationale management in software engineering.

2.1 Introduction

2.1.1 Rationale for Software Artifacts vs. Rationale for Physical Artifacts

Rationale research and applications have been conducted not only in software engineering (SE), but also in a variety of other fields, including mechanical engineering, civil engineering, architecture (building design), architectural engineering, urban design, city planning and policy making. Almost all of these fields deal with the creation of physical artifacts, such as machines, bridges, buildings, and cities.

Research on rationale for decision making began in architecture (building design) and urban planning with Rittel's work on IBIS (Kunz and Rittel 1970). The initial adoption of this work for use in software engineering (Conklin and Begeman 1988; Potts and Bruns 1988) was based on the notion that there are crucial commonalities between the processes of creating software and the processes of creating physical artifacts like buildings and cities. This notion derives some additional plausibility from the fact that design patterns (Gamma et al. 1995), which have found such widespread acceptance in SE, were also originally invented for use in creating buildings and cities (Alexander et al. 1977). In Chapter 4 we discuss what software engineering can still learn from work on rationale for physical artifacts. In the current chapter, however, we concentrate on the differences between the task of devising effective rationale management for SE and the comparable task in physical artifact development. In particular, we argue that these differences are crucial for the success of rationale-based software engineering for two reasons. One is that software engineering has unique problems that can be alleviated with rationale management. The other is that it also has unique advantages that aid the implementation of rationale management.

2.1.2 Objectives of this Chapter

The central objective of the chapter is to point out important differences between software development and the development of physical artifacts, i.e. differences that have crucial significance for making rationale management a practical reality in software development. Section 2.2 describes the special roles of the computer in software engineering and how these create unique opportunities for effective rationale management in software development. Section 2.3 looks at the role of iteration in software development and how this differs decisively from its role in the development of most physical artifacts. This difference creates opportunities for rationale management in software development but it also calls for approaches to rationale management that go beyond those created for physical artifacts. Finally, Section 2.4 summarizes the ways in which software development is different and the significance of this fact for rationale-based software engineering.

2.2 The Roles of the Computer

Where the activities of an artifact's lifecycle involve using computers, it becomes possible to employ rationale management software to greatly facilitate the capture, editing, structuring and retrieval of rationale. This is especially true when rationale management functionality is integrated into software used in decision making, such as CAD systems (Fischer et al. 1996), CASE systems (Oinas-Kukkonen 1988) and programming IDEs (Burge and Brown 2006) as well as systems for computer-supported collaborative work (McCall and Johnson 1997). There appears to be a broad consensus that without software support, rationale management is generally infeasible except for very small projects. As a consequence, in complex, real-world projects the question of whether a lifecycle activity is computer supported becomes decisive for determining the viability of rationale management in conjunction with that activity. The sections that follow argue that in this respect SE has dramatic advantages over fields that aim the creation of physical artifacts.

2.2.1 Comparison of the Roles of the Computer in the Lifecycles of Physical and Software Artifacts

The development processes for physical and software artifacts have a number of important similarities. Both involve the identification and analysis of requirements as well as design processes that center on the creation of models of the artifact being developed. In both domains, the computer has gained an increasingly important role in supporting design and requirement-related processes. And it is plausible in both cases that in the future all aspects of these processes might come to be computer supported.

In those aspects of the artifact lifecycle that are not related to requirements or design there are profound differences between the roles of the computer with respect to software and physical artifacts. These differences derive from the simple fact that physical artifacts do not require computers for their existence while software artifacts do. Though computers can in certain cases play a role in the *construction* of physical artifacts, as with CAD-CAM (computer-aided design computer-aided manufacturing), computers are not necessary for the construction of most physical artifacts. Furthermore, they are never sufficient for the construction of any physical artifacts because non-digital, physical means must always be employed. Functioning software artifacts, however, cannot be constructed without the use of computers. In fact, use of computers is both necessary and sufficient for the construction of software. We might write code using pencil and paper, but this code does not become software until it can be used by computers.

The role of the computer also differs in the *use* of physical and software artifacts. While there do exist some computer-mediated ways of using physical artifacts, such as by means of tele-robotic systems that move around or through physical artifacts and make it possible for human users to manipulate them. Using artifacts in this computer-mediated way is, of course, rare. Almost invariably, the use of physical objects is by purely non-digital, physical means. Software artifacts, by contrast, can only be used through use of computers.

The differences in construction and use in turn imply differences in the roles of computers in maintenance and testing. After all, maintenance involves construction and testing involves at least simulated use. Maintenance of physical objects necessarily involves physical means. And in fact, the use of computers in testing and maintenance of physical artifacts is still relatively uncommon and seems likely to remain so for many years to come. With software, however, neither maintenance nor testing of constructed artifacts is possible, or even conceivable, without the use of computers. The result is that the computer is invariably much more heavily used in the maintenance and testing of software than in the maintenance and testing of physical artifacts.

In summary, when it comes to construction, testing, maintenance and use of artifacts, there is a profound difference between the roles that computers play with respect to physical artifacts and software artifacts. Computers are ubiquitous in these aspects of the software lifecycle, yet they are rarely used in these aspects of the lifecycle of physical artifacts. If this ubiquity is combined with the increasing role of computers in software requirements engineering and design, we see that the computer will eventually be ubiquitous in all aspects of the software lifecycle, i.e. in all aspects of SE.

The intent here is not to downplay or to diminish the importance of the computer in the development of physical artifacts, but merely to point out that the role of the computer in the software lifecycle goes substantially beyond the role of the computer in the lifecycle of physical artifacts. Software differs from physical artifacts in having the computer as a *common medium* for *every aspect of its creation and existence*, including design, implementation, testing, use, and maintenance.

2.2.2 The Significance for Rationale Management in Software Engineering

2.2.2.1 General Implications for Support of Software Engineering Rationale

The differences between the roles of computers in the respective lifecycles of physical and software artifacts have crucial implications for support of rationale. The greater role of computers in the construction, testing, use and maintenance of software means that there are greater possibilities of using rationale in these activities. In other words, SE has the potential for using rationale in more lifecycle activities than do fields that specialize in the development of physical artifacts. This potential has two parts: many more places where rationale can be captured and many more places where it can be used to improve the artifact creation process.

The ubiquity of the computer in the software lifecycle has important implications for rationale-based software engineering, though research on some of these implications is still in the early stages. One such implication is the potential for extending rationale research beyond the current focus on design and requirements engineering to other activities in the software lifecycle. Another implication of this ubiquity is the potential for computer-mediated communication, collaboration and participation involving participants in development, maintenance and use of software. Such communication provides a rich source for the capture of rationale (Shipman and McCall 1997) as well as an excellent vehicle for the delivery of rationale to those who need it. This sort of communication can provide feedback from construction and use that informs iterative and incremental approaches to design and requirements engineering. It can also provide valuable feedforward about requirements and design intent that informs construction, maintenance and use.

2.2.2.2 Linking Software Artifacts to their Rationale

One important consequence of the ubiquity of the computer in the software lifecycle is that the rationale that is about some feature or characteristic of the system can be directly linked to the part of the software that implements that feature or characteristic. This has a number of implications for rationale management. One is that the artifact can be used as a way of rapidly accessing rationale. For example, Burge uses source code in the Eclipse IDE to alert programmers to the existence of rationale about individual pieces of code (Burge and Brown 2006). This is, of course, made possible by augmenting Eclipse with rationale management functionality. With this sort of augmented IDE, the artifact can in effect be used as a way of indexing rationale. Though Burge also uses her RATSpeak schema for structuring rationale, some, such as Schneider (2006), use linkage of rationale to software as a substitute for use of a schema. This has the potential of eliminating the need for a schema, which in turn has the potential of dramatically reducing the amount of work required for capture of rationale—or at least the rationale associated with construction decisions.

Yet another potential value of linking rationale to software is that it provides a basis for checking whether the decisions about the requirements, design and implementation correspond to the as-built state of the code. Such checking might even be partially or completely automated. Automated checking would require that the computer be able to "understand" the denotation of the textual descriptions of decisions in the rationale for requirements, design and implementation.

It might be argued that the linking of the rationale to the artifact is not really something that makes software development different from the development of physical artifacts, because with CAD systems, rationale could be linked to the digital model of the physical artifact. A crucial difference, however, arises when the artifact is *actually constructed and put into use*. A physical artifact has no intrinsic ability to have rationale linked to it—since the rationale is digital and the artifact is not. Of course, someone might devise a way of linking rationale to various parts of a physical artifact—using, RFIDs, bar codes or some as-yet-unknown technology. The crucial point, however, is that such linking is inherently much simpler and easier to accomplish with software than with a building or some other physical artifact.

The relative ease of linking rationale to software artifacts implies that a person constructing, modifying or re-using software could easily have access to the rationale behind the design and the requirements as well as the rationale of others who have worked on the system—and, in fact, to the rationale of every stakeholder associated with the system's creation, revision and use. Furthermore, if the links to rationale are preserved when the code is compiled, then even users could have access to rationale for the software. In fact, Haynes (2006) has advocated and experimented with using design rationale that is linked to compiled software to explain the functionality of complex systems to their users.

2.2.2.3 Using Networked Computers to Capture and Retrieve Stakeholders' Rationale

The fact that the networked computer is the ubiquitous platform for every activity in the software lifecycle, including implementation and use of the software, means that all the stakeholders for a software project, including the developers, maintainers and users, could in principle input their own rationale at any time and retrieving the rationale created by any and all other stakeholders at any time. Rationale methods for doing this already exist. Carroll's Scenario-Claims Analysis approach is already well suited to capturing user evaluations of system features in the context of use. And the decision-centered approaches to argumentative rationale, including IBIS, PHI, QOC, DRL and RATspeak are well-suited to documenting the rationale of the development and maintenance teams. What needs to be done is to create software systems that can support this full spectrum of rationale management. This will require a great deal of work, but the potential is there. No type of physical artifact development offers such potential.

In fact, if communication amongst the stakeholders is integrated into the software environments for development, maintenance and use, such communication could be automatically captured and used as a basis for the system's rationale. This by itself would go a long way towards solving one crucial part of "the capture problem," which has been the main obstacle to practical use of rationale management—that part being the recording of substantial amounts of rationale. In fact, in addition to acquiring large quantities of rationale, it would enable acquiring rationale from the entire spectrum of stakeholders in the project. The advantages this would offer for collaboration, coordination and project management can hardly be overestimated.

The other part of "the capture problem" actually has nothing to do with capture per se. It is the problem of structuring and indexing captured rationale. Without this, the rationale cannot be effectively retrieved when needed. This part of the problem might well be solved with the help of the sort of artifact-based structuring described above.

Implementing a practical system that all stakeholders would actually use for capturing, structuring, indexing and retrieving rationale would be an extremely ambitious task, and one that might be still more complex than the above-given description suggests. But it is possible for software engineering in a way that simply does not exist for other fields that seek to aid the development of physical artifacts. This fact alone suggests that there is more hope for success in creating a truly rationale-based software engineering than for implementing a rationale-based approach to any field of physical artifact development.

With physical artifacts, the rationale for their development ultimately becomes disconnected from that artifact, which means there is a natural tendency for that rationale to become inaccessible to people who use these artifacts or try to learn from their development. With software there exists the possibility for rationale to be permanently bound to the software it discusses and thus available to all who have access to the software regardless of how much time has passed. This would facilitate both the reuse of the software and the design of similar software in the future.

The permanent connection of software to its rationale would facilitate learning from previous projects and even the development of cumulative stores of rationale that are not only added to over many years, but that are also progressively refined in the detail, completeness, consistency, organization and indexing as they mature. This learning would, of course, be greatly enhanced if the rationale included extensive feedback from users, so that future developers can judge whether the expectations of the developers about the quality of the software were matched by the experiences of the users. Such stores of rationale could serve as organizational memories for project teams, companies and the software engineering field as a whole.

2.3 Iteration in Development

2.3.1 The Role of Iteration in Different Types of Development

Iteration has historically been the subject of radically different points of view in software engineering. Early on, many campaigned against it; more recently, many have campaigned for it under such various labels as "incremental," "evolutionary," "agile" and, of course, "iterative" development. One does not have to take sides in this controversy to recognize that iteration is possible in software development to an extent and in ways not found in the development of most other kinds of artifacts, especially larger-scale physical artifacts like buildings and cities. In fact, it is precisely the *manifest possibility* of iterative approaches to software development that has enabled it to be the subject of controversy amongst software engineers.

Independent of the iterative approaches currently being promoted in SE, it is clear that iteration, in the sense of repeatedly learning from experience with constructed systems, is already deeply rooted in conventional software development. This is especially clear in the cases of COTS (commercial off-the-shelf) and open-source software. Such "products" generally have great longevity and go through many different release versions, each of which may be preceded by alpha and beta versions. Their development is characterized many iterations of redesign, recoding and patching, all informed by feedback from constructers, users and maintainers of previous versions.

If, however, we look at the role of iteration in the development of large-scale physical artifacts such as buildings a dramatically different picture emerges. While there is good deal of iteration in the *virtual world* of CAD models, there is relatively little iteration in the *real world* of the constructed artifact itself—especially in comparison with SE. One important reason for this appears to be that the costs of changes in construction would be excessive as a fraction of the overall cost of the project. Another reason is that an incremental approach in which large artifacts, such as buildings, are partially constructed, then inhabited and then later refined seems to be both dangerous and infeasible for a variety of reasons. For example, while people can continue using an old software version while a new one is being constructed, constructing a new version of a building would probably require evacuation of the building. Thus, the concept of *version*, in the sense of a constructed artifact, plays no significant role.

Like version, the concept *prototype* also has little or no role in the development of large physical artifacts. The only real use for prototypes, in the sense of full-scale, functioning artifacts, is for possible testing of a few small parts of the final design—for example, a prototype of a wall panel for a building being created. Full-scale, usable prototypes of a building are generally out of the question; they are simply too expensive. These facts rule out *iterative development* of buildings, in the sense in which this term is used in software engineering.

2.3.2 Implications of Iteration for Rationale Management in Software Engineering

2.3.2.1 The Importance of Rationale in Iterative Development

Each iteration in an iterative development process introduces change into the software system. At a minimum there is change in the construction of the system, but there may also be changes in the design and even in the requirements. Before any such change is made it is important to know what the rationale was for the previous state of the system, because this can help to avoid breaking or losing functionality when changes are implemented. Records of the dependencies between design decisions are especially important because they enable developers both to *predict* the consequences of changes and to *cope* with these consequences. If developers do not have a solid knowledge of rationale, there is the danger that the software will, over many iterations, "drift" and lose the integral unity of its design and construction. It might then "grow like topsy," becoming progressively more disorganized and less robust.

2.3.2.2 The Absence of Explicit Iteration in Most Approaches to Rationale

Both the possibility and current popularity of iteration (Rajlich 2006) in software development make it important to ask what its implications might be for rationale management in software engineering. Unfortunately, almost all argumentative approaches to rationale management in software engineering—including IBIS, PHI, QOC, DRL and RATspeak—fail to provide any explicit account of how they would deal with iterative development. These approaches base decision making on a purely verbal process of argumentative deliberation. There is no explicit role for action, construction, versions, prototypes, testing, use, experience or empirical evidence. This, however, does not necessarily mean that they cannot be used in iterative software development. It merely means that they have not yet indicated how they would do so. In fact, it seems that lessons learned from experience with constructed and released software could easily be incorporated into argumentation; but there has been little discussion in the literature about how this would work. Until more literature is generated on this subject, it will up to each development team to determine how to use these rationale methods in the context of iterative software development. However, Chapter 16 of this book, entitled, "A Conceptual Framework", however, attempts to partially remedy this problem by giving a theoretical account of how rationale might support iterative software development.

There are some approaches to rationale that current deal with iteration explicitly. One of these is Scenario-Claims Analysis (SCA), which Carroll and Rosson (1996) devised explicitly to support what they describe as "deliberated evolution" in "the task-artifact cycle." In addition, the Win-Win rationale method is explicitly tied to Boehm's Spiral Model of software development (Boehm and Kitapci 2006).

2.3.2.3 Rationale as a Means for Benefiting from Lessons Learned

If software development is an inherently iterative process in which software is improved through experience with constructed systems, then it is largely *a process of learning*. One crucial opportunity for improving the quality of future software development is to make sure that the hard-won lessons learned from iterative development efforts are available for future development efforts. Documenting rationale behind current development provides a means for doing this. It is crucial not merely to document the reasons for the decisions taken but also the arguments against those decisions, the alternatives to the decisions and the argumentation on these. Without this sort of documentation it is easy to fall into the trap of being seduced by intuitive but mistaken solution ideas. Without documentation of why bad ideas are bad, we doom future generations of developers to find out the hard way that these ideas are bad.

If future developers can learn the lessons of previous developers without having to repeat their experiences, it would seem that the amount of iteration required in development might be reduced. Only time and experience with documented rationale will tell if this is the case.

To maximize the benefits of lessons from the past, we need long-term storage of rationale and widespread access to that rationale. Ideally, we need cumulative stores of rationale that grow and evolve through use. A number of approaches are possible, including rationale centered on design patterns (Pena-Mora and Vadhavkar 1996; Hagge et al. 2006), domain-oriented issue-bases such those used by JANUS (Fischer et al. 1996) and PHIDIAS (McCall et al. 1992), and approaches based on Case-Based Reasoning, such as that used in the ARCHIE system (Kolodner 1993) (see Chapter 4 of this book, "Learning from Rationale Research in Other Domains").

2.4 Summary and Conclusion

Rationale research began in fields that dealt with the development of large-scale physical artifacts, like buildings and cities. The development of software differs from the development of such physical artifacts in ways that are crucial for the success of rationale management in software engineering. The ubiquitous role of the computer in every aspect of the software lifecycle give software engineering a potential to capture rationale from every type of stakeholder in a development project and to enable each of those stakeholders to retrieve rationale from every other stakeholder. No comparable potential exists in physical artifact development. In addition, because the software artifact is constructed and used on the computer, the rationale from stakeholders can be linked to the sections of the software, thus enhancing retrieval and potentially easing the work of capturing, structuring and indexing rationale.

Finally, the inherently iterative nature of much modern software development creates both challenges and opportunities for software engineers. The challenges include modifying rationale methods to reflect the iterative reasoning processes in this development. The opportunities include the capability of learning from past development efforts through the building of cumulative stores of rationale that grow and evolve through use in the context of development.

3 Rationale and Software Engineering

Software engineering, the process of developing software intensive systems, is a complex area. This chapter introduces software engineering as well as the potential benefits in capturing, maintaining, and reusing rationale to support it.

3.1 Introduction

3.1.1 Software Engineering

According to the IEEE (IEEE 1993), software engineering is "the application of a systematic, disciplined, quantifiable approach to the development, operation, and maintenance of software; that is, the application of engineering to software." A more detailed definition of Software Engineering, and the one that we use during this book, was provided by Finkelstein and Kramer (2000):

> SE focuses on: the real-world goals for, services provided by, and constraints on such systems; the precise specification of system structure and behavior, and the implementation of these specifications; the activities required in order to develop an assurance that specifications and real-world goals have been met; the evolution of such systems over time and across system families. It is also concerned with the processes, methods and tools for the development of software intensive systems in an economic and timely manner.

Both the IEEE definition and the Finkelstein and Kramer definitions stress the necessity for a disciplined process of software development. This discipline is what puts the "engineering" in software engineering.

3.1.2 Software Engineering Rationale

Much of the research on rationale has addressed Design Rationale (DR). In domains such as engineering design, most critical decisions are made at design time and that is when the majority of the rationale is captured and used. During software development, while most development methodologies include a phase called design, decisions that drive software development are made all throughout the process. We therefore view rationale as something that can be captured and used at all stages. In this book we use the term *software engineering rationale* (SER) to encompass all different types of rationale in many SE processes (Dutoit et al. 2006) and to serve as a base for examining how SER can support the entire software engineering process.

3.1.3 Objectives of this Chapter

This chapter begins with a description of how rationale can be used to help define and implement the software process. This is then followed by a description of how rationale can support project management. The remainder of the chapter introduces how and when rationale can be used in software development.

3.2 Rationale and the Software Process

3.2.1 Software Process Definition and Implementation

In order for software development to be performed in a systematic and disciplined approach, it is necessary to follow some defined software engineering process. There is no single software development process that fits all types of software development. Instead, the software process used should be chosen, or defined, to best meet the organizational needs of the software developers as well as any process requirements that may be mandated by the client. According to the IEEE SWEBOK (Software Engineering Body of Knowledge) (IEEE 2004), software engineering process (SEP) definition/development can be broken into four sub-areas: 1) Process Implementation and Change, 2) Process Definition, 3) Process Assessment, and 4) Process and Product Measurement.

The Process Implementation and Change sub-area defines what needs to be known in order to either implement a new software engineering process or to change an existing one. This includes definition of the infrastructure needed for process management, determining how the process will be managed, and selecting an appropriate quality improvement model.

The Process Definition sub-area involves selecting the appropriate software life-cycle model, the software life-cycle process, determining the appropriate notation to describe the software process, adapting the selected process to meet the needs of the specific organization, and determining how, or what portions of, the process can be automated using process support tools.

The Process Assessment sub-area utilizes assessment models. The Capability Maturity Model (CMM) (SEI 1997) and CMMI (CMMI 2006) are two examples. Process assessment also requires process assessment methods that can use information about the process to give it a rating, or "score."

Finally, the Process and Product Measurement sub-area describes the need to measure process outcomes (its success at meeting process outcomes) and to perform product measurement to look at its size, structure, and quality. Of course, deciding what data to collect is insufficient, it is also necessary to decide how to determine how to assess the quality of the measurement results. A rigorous quality improvement process also involves collecting measurement data over time into a repository, modeling the information, and determining how the information can be fed back into the process on future projects.

3.2.2 Rationale and SE Process Decision-Making

We will describe the role of rationale in the software life-cycle and in software process improvement later in the book. Here, we will address how rationale can support the process definition process described in the SWEBOK as outlined above. Determining what the appropriate software development process is, and how that process should be managed and measured, involves making a number of very crucial decisions. The decision-making process involves determining what are the software process goals, what are the alternative means for achieving those goals, and evaluating those means to determine which alternatives best suit the goals of the specific organization and project.

Process Implementation and Change. In order to implement a new process, or change an existing one, there are many decisions that must be made. What are the requirements for the new/adapted process? What changes should be made to the current process infrastructure? How is the process going to be managed? Which quality improvement model best suits the needs of the project? The rationale for the choices made when making these decisions can be used to determine if the reasons for these choices is consistent with project goals. It can also be compared with that from prior projects to see where past processes can be re-used or adapted to meet new process needs.

Process Definition. The choice of process, and how rigorous that process should be, will have a significant impact on the software project. There are tradeoffs that need to be made between having a well defined and rigorously monitored process and the cost and time that may entail. Software life-cycle models are not "one size fits all." Selecting the appropriate model for a specific project involves careful examination of alternative life-cycles and their advantages and disadvantages relative to the needs of the organization. There are also many choices that need to be made when deciding if, and how, the process requires adaptation to meet specific organizational goals. It is important that adaptations are consistent with the goals of the life-cycle and to not counter its advantages. Process automation is also an area where decisions must be made. If automation is a high priority, it may prove to be a key driver in selecting the software process. The process may be chosen based on the tool support available and what that tool support is likely to cost.

Rationale should be recorded for the reasons behind the choices made. The explicit articulation of tradeoffs made will ensure that the choices are made for the right reasons and, if these decisions are revisited for future development efforts, that the effort that went into making these crucial decisions can be assist in making the correct decisions in the future.

Process Assessment. The choice of how a process will be assessed may or may not be under the control of the software development organization. In either case, the standards used to evaluate the software process can be captured in the rationale as criteria used to assess the other decisions made during the process definition process. The process outcomes identified will be the main criteria used to determine the process measurement strategy.

Process Measurement. There are many aspects to the software process and product that *could* measured during development. The question is, which of these *should* be measured in order to assess the software projects success at achieving process outcomes. Again, there are tradeoffs made between the time and effort it takes to perform process and product measurement against the value of the information obtained. Choices may be made based on tool support available to assist in this effort.

Capturing the rationale for these decisions can help to clarify what measurement options should be considered and what the reasons are for choosing them. The knowledge captured in the form of rationale can also assist future projects when they need to make similar decisions.

3.3 Rationale and Project Management

The Project Management Institute defines project management as "the application of knowledge, skills, tools, and techniques to project activities to meet project requirements" (PMBOK 2003). This definition is rather general but it is commonly understood that good project management is essential to ensuring that a project meets its goals of delivering quality software on time and within budget. Management needs to work successfully with the client to ensure that their needs are understood and met while also working with the developers to ensure that they have the knowledge and resources necessary to successfully develop the software product.

As in software development, rationale can play multiple roles. Rationale can assist with guiding and capturing the decision-making process when developing the management strategy for a project. As with software development processes, there is not one management solution that will work under all circumstances. Processes used in the past require tailoring to meet the needs of specific projects and the skills of specific teams. Rationale captured for management choices in the past can be used to determine if those choices are still valid for future projects.

Examples of some management choices include:

- Status reporting requirements for project teams
- Project team structure (size, distribution of responsibility, communication strategy)
- Necessity of hiring consultants with key technical expertise
- Frequency and duration of status meetings
- Role of software tools in the software project

Criteria for making these choices might include:

- Team member expertise and experience
- Team familiarity—experience of team members with each other
- Value of permanent employees learning new technology for future projects
- Budget provided for tool aquisition
- Management experience
- Customer flexibility (in both deliverables and schedule)

It is critical that these key management decisions be made based on an understanding of the criteria that impact their success. Using the rationale to capture and evaluate these decisions helps to ensure that the management strategy selected best suits the needs of the client, product, and team.

Rationale can also assist with many project management or project management related tasks. Charette (1996) states that "large project management is risk management." The identification of risks is a crucial factor in successful software development. Capturing these risks, alternative mitigation strategies proposed, and the mitigation strategy used, serves to both clarify the risk management process on the current project as well as form a knowledge base of risks, strategies, and outcomes for use on future projects.

Another aspect of software development where project management plays a key role is in the reconciliation of stakeholder viewpoints. Theory-W (Boehm and Ross 1989) is a software project management theory where the main goal is to "make everyone a winner." Theory W is based on Fisher and Ury's (1981) negotiation approach, where a key part of the negotiation involves identifying options and evaluating those using objective criteria. In Theory W, the key to a successful negotiation is to identify the stakeholder win conditions and to find options that create the win-win situations. The generation of these options and win-conditions is supported using the WinWin support system (Boehm et al. 1995). The information captured in WinWin is, in essence, the rationale behind the software requirements (Boehm and Kitapci 2006).

One of the more successful uses of argumentation-based rationale is to assist with structuring discussion during project meetings. The Issue Based Information System (IBIS) notation (Kunz and Rittel 1970) is the basis several of several systems applied to capture discussions in meetings. The itIBIS (indented text IBIS) system was used at NCR to capture project team meetings (Conklin and Burgess-Yakemovic 1996). This helped to focus discussion and point out potential problems with the requirements. Converting the textual rationale into a graphical form (gIBIS) exposed several problems with the proposed design that would probably not have been detected otherwise. The use of IBIS to aid in collaboration has continued with the Compendium project (Buckingham Shum, et al. 2006) to perform "Dialogue Mapping." In their approach, a trained facilitator uses Compendium to capture discussion in an IBIS format during meetings. The results of the discussion can be displayed in real-time to allow meeting participants to view, and reflect on, the discussion taking place.

3.4 Rationale and Software Development

The previous sections, we described how rationale can assist with defining the software development process and in managing the implementation of that process. Here, we highlight uses of rationale during the software development process by describing why rationale is needed, what some of the uses of rationale are, when rationale can be used during the process, and finally how it can be used. These areas are our primary focus during the remaining chapters of this book.

3.4.1 Why Capture Software Engineering Rationale?

Earlier in this book we defined rationale and its importance in software engineering. The success of any software project is dependent on the right choices being made during its development.

Software engineering contains many key challenges that can be addressed by the capture and use of rationale:

- *Software system longevity.* Software systems have been shown to remain in operation longer than the original developers probaby anticipated. This longevity, and the need to continualy evolve software to keep it viable, means that it is essential to understand the reasons behind development decisions made years earlier.
- *Iterative nature of software development.* Many current software development processes utilize some form of iteration in order to increase their ability to adapt to changing requirements and technology. As development progresses, criteria appearing in the later iterations may affect decisions made in the earlier ones. The rationale can help to assess the impact of the changing criteria and guide the developer in making changes that implement the new functionality with minimum risk to that implemented earlier.
- *Stakeholder involvement.* There are many different stakeholders in a software development effort who have their own, and sometimes conflicting, goals for the system. For example, the customer is concerned with the functionality provided by the system; the end user is concerned with how well it helps them perform their tasks and how easy it is to learn and use; the developers are concerned with how difficult it will be to implement; the managers are concerned with how long implementation will take and how much it will cost; all stakeholders are concerned with the reliability of the delivered system; etc. Capturing the decision-making process, and the stakeholders having input into that

process, can serve as a basis for negotiation. Rationale also captures how the different stakeholder priorities affect the developed system.

- *Knowledge transfer*. Significant amounts of expert knowledge are involved in the development of a large software system. This is information that will be lost if not documented, particularly at times of high turnover in the software industry. Rationale can serve as a key component to an organizations knowledge management strategy.
- *Increasing size and complexity of software systems*. Software systems have long since passed the point where their design is simple enough to exist in the heads of their developers. Rationale can assist as a "memory aid" to assist developers in remembering why they made their earlier decisions. Rationale can also be used to index into the code and documentation to determine the impact of changing decisions on the software.

3.4.2 What are the Uses of Software Engineering Rationale?

In order to convince software developers that capturing rationale is worth their time and effort (and convincing software managers that capturing rationale is worth some additional up-front costs), it is essential that the rationale is useful both during the initial requirements and design stages and later as the software is maintained and reused. We have identified several key areas of rationale use:

- *Presentation*. The use that immediately comes to mind for rationale is its ability to document the decision-making process. The ability to browse through, or query, the rationale-base to learn more about the decisions can assist developers in learning about the software, preventing the duplication of past work, and avoiding errors. The usefulness of the presented rationale will be dependent on the method of presentation. Ideally, presentation should be done within the same tools that are already in use to develop the software. The developer will be far more likely to know the rationale is available and take it into account when making decisions if they do not need to use an additional tool.
- *Evaluation*. The CMMI (CMMI 2006) Decision Analysis and Resolution (DAR) process area stesses the importance of performing a "formal evaluation" of selected issues by evaluating alternative solutions (that address those issues) against criteria. Rationale can support this type of calculation by providing detailed information about the solution alternatives and their relationship to the decision criteria (such as requirements, quality attributes, and assumptions). This information can be

used to rate or rank the alternatives to evaluate the quality of the decision results. Rationale also supports usability evaluation, as demonstrated by the Scenario-Claims Analysis approach (SCA) (Carroll and Rosson 1992).

- *Collaboration.* Later in this book we describe how software development is almost always collaborative work. Rationale's importance to collaboration during software engineering was highlighted in Jim Whitehead's talk as part of the Future of Software Engineering track of the 2007 International Conference on Software Engineering (Whitehead 2007). Whitehead views architecture and design as "argumentative proceses" and proposes rationale capture, in the form of "collaborative argumentation" as an effective means of supporting these processes. The ability for rationale to support and capture this the negotiation required during software development has been demonstrated by many approaches, such as the WinWin (Boehm et al. 1995) and Compendium (2006) systems described earlier.

- *Change Analysis.* As mentioned earlier, software development is an iterative process. Software requires change both during the development process, as more information is learned about the requirements and incorporated into the software, and afterwords as it enters the maintenance and evolution stage of its life. Software may require changing for a multitute of reasons but one thing remains certain—the need to understand how the proposed changes impact the existing software. This includes both determining where the changes need to be made and also how those changes may affect the ability of the software to meet the requirements, quality criteria, etc. that were the basis of the decisions made during its initial development. With appropriate tool support, rationale can be used to identify change location and change impact. Rationale-based consistency checking can aid in consistency management—an ongoing process during software development and maintenance.

3.4.3 When Can Software Engineering Rationale be Used in Software Development?

As mentioned earlier, rationale can support many aspects of software development and is not constrained to the design stage. These aspects include the "standard" development stages of requirements, design, etc. and also the cross-cutting areas of project management and re-use.

- *The Software Lifecycle.* Rationale can play a role in any of the software lifecycles selected to guide the software development process. Rationale also has a role in software process improvement, as mentioned earlier in this chapter.
- *Requirements Engineering.* Rationale is involved in software requirements in several ways. One is in requirements elicitaiton and documentation. The rationale is a natural place to capture the relationship between the software requirements captured during elicitation and the source of those requirements. This provides a "rich traceability" back to the original customer requirements (Dick 2005; Hull et al. 2002). As with all aspects of software development, negotiation plays a role in requirements engineering as all stakeholders need to agree on what the requirements are. This negotiation and the parties involved can be captured in the requirements rationale. Requirements also appear in the rationale for the system as the arguments for and against alternatives. Capturing this information, and associating it with the code that implements the alternatives, is a form of requirements traceability (Burge and Brown 2007).
- *Software Design.* Since much rationale research has been in the area of design rationale, it is no surprise that rationale for software design, and more specifically software architecture, is an active research area. Software architecture, while traditionally thought of in terms of components and connectors, is seen by some as "a composition of architectural design decisions" (Bosch 2004). This decision-centric view has encouraged more research into capturing the knowledge behind those decisions, as shown by workshops such as the SHaring and Reusing Architectural Knowledge (SHARK).
- *Software Verification and Validation.* This is an area where the capture and use of rationale remains largely unexplored. Still, decision-making in software engineering does not stop when the development is complete. The planning and execution of an effective testing strategy requires making complex tradeoffs between cost and quality in order to ensure that the software meets the needs of its users while keeping testing costs under control. Rationale for the choices made when selecting testing methodologies and tools should be captured so that it will be available for use by subsequent projects or if the testing strategy of the current project requires re-evaluation.
- *Software Maintenance.* One of the areas where the availability of rationale can be most valuable is during software maintenance. The challenge of software maintenance is ensuring that software evolves without damage to, or reduction in, functionality needed by its users.

This is difficult because the maintainers may not be the same people who initially developed the code and who have an often steep learning curve to understand an unfamiliar piece of software. The ablity to utilize the past experience of software developers via access to their rationale supports these goals.

- *Software Re-use*. Re-use has often been refered to as "the holy grail" of software engineering. The ability to re-use software systems or components has shown great promise in allowing software delivery with fewer defects, higher quality, and in significantly less time. There are many types and levels of software re-use and while all have advantages, re-use is not without its risk. Rationale can play several roles in re-use. One is to support decision-making on if and when re-use is appropriate for any given project. There may be some cases where the risk outweights the benefits. Another use is to capture the reasons behind the decisions on what should be re-used. There may be several re-use alternatives that should be examined. Rationale can also be used to evaluate re-use candidates. If the rationale behind those candidates is available, this information would provide valuable insight into the decisions that went into their design.

3.4.4 How Can We Support Software Engineering Rationale Use in Software Development?

In order for Rationale-Based Software Engineering to live up to its promise, we need to develop Rationale Management Systems that support its capture and use. As in any software development project, the first step is to identify the requirements. What are the uses of rationale that such a system needs to support? How does rationale, as we currently understand it, support software engineering and when does it fall short? How do we address those shortcomings?

Later in this book we provide two frameworks, one defining the key concepts in Rationale-Based Software Engineering and their relationships (the Conceptual Framework) and one that provides a framework for RMS development that supports the key features of RBSE needed to support software development (the Architectural Framework).

3.5 Summary and Conclusions

Rationale can play many roles throughout the software development process, both descriptive - by providing a richer view into the decision-making process, and prescriptive – by guiding that process and evaluating its results. There is however a small literature of doom-and-gloom discussions that dismiss the value of rationale relative to its cost, some even implying that the additional cost could make the difference between software project success or failure (Grudin 1996). Cost is an important factor in the equation, but it not a simple linear factor. Indeed, most nihilistic accounts of rationale describe development projects where rationale practices were implemented narrowly, manually, and incompletely.

Rationale provides technical leverage throughout all the processes and activities of software development. A broad approach to capture and reuse of rationale is required to enjoy multiplicative benefits of pervasive rationale practices. Software tools to support partial automation of rationale management can reduce the cost side of the equation even further. Finally, implementing rationale practices thoroughly in development organizations is critical. Process improvement efforts such as the CMM and CMMI involve rigorous documentation of software development that takes both time and effort. Initial studies on the CMMI (Goldenson and Gibson 2003) show that many of the companies studied showed cost, schedule, and quality improvement after adopting the processes.

When rationale practices are adopted broadly and with appropriate tool support, and when they are adopted thoroughly in development organizations, rationale has the potential to yield benefits that far outweigh its costs.

4 Learning from Rationale Research in Other Domains

While issues of rationale usage in software engineering (SE) often differ crucially from those of rationale usage in other domains, there is still the possibility of learning a great deal from research on other domains. This is suggested by the fact that rationale research in SE originally derived from Rittel's much earlier rationale research in architecture (building design), urban planning and policy making. In addition to this work, which is still not widely known in SE circles, there is research on rationale that has been going on in various engineering disciplines for as much as 20 years. All of this work provides potentially valuable lessons for SE researchers and developers. This chapter will look at some examples of this work that could have important implications for rationale research in SE.

4.1 Introduction

4.1.1 Research on Rationale in other Domains

Research on design rationale began with Rittel's Issue-Based Information System (IBIS) (Kunz and Rittel 1970) and its applications to urban planning, architecture (building design) and governmental policy making in the 1970s and 1980s. By the late 1980s software engineers at the Microelectronics and Computer Technology Corporation (MCC) were looking at adapting Rittel's method to their own field and developing appropriate computer support (Conklin and Begeman 1987; Potts and Bruns 1988). Since then many other researchers involved with software engineering (SE), human-computer interaction (HCI) and other software-related related fields have created various rationale approaches, including QOC (MacLean, Young and Moran 1989), DRL (Lee 1991), RATSpeak (Burge and Brown 2004) and many others. Most of these approaches continue the basic tradition started by Rittel, while suggested various modifications meant to go beyond Rittel's IBIS and better fit rationale to the SE domain.

Chapter 2 of this book emphasizes that there are some crucial differences between the problems of rationale usage in the SE domain and rationale usage in the domains of physical artifact creation. At the same time, there continues to be a considerable overlap in the issues facing rationale researchers in these two types of domains. This suggests that researchers in these domain types might still have much to learn from each other. This chapter explores this topic by presenting some examples of rationale research in design and engineering.

4.1.2 Objectives of this Chapter

Rather than attempting a comprehensive survey of rationale research in other domains, this chapter will concentrate on examples of such research that raise important issues for research on rationale support in SE. For these examples, the issues raised mostly have to do with the way in which they use computers to support rationale; therefore, this chapter will go into more detail on the rationale management software systems than is generally the case in the remainder of the book.

The approaches and systems described in the chapter all deal with the rationale for design. For the examples discussed, this chapter will first identify crucial functionality that they bring to the support of rationale, functionality not currently found in rationale management systems for SE. Connections to existing research on software engineering rationale (SER) will then be identified. The potential advantages of adopting this functionality in SER support systems will then be discussed; and potential challenges to implementing this functionality in SE will be described.

4.2 Domain-Oriented Design Environments Using PHI

4.2.1 PHIDIAS and JANUS

The PHIDIAS (PHI-based Design Intelligence Augmentation System) project (McCall et al. 1990) began in 1985 with the goal of adding a CAD subsystem to the MIKROPLIS hypertext software. MIKROPLIS (McCall et al. 1981; McCall 1991) was a hypertext authoring system devised in the early 1980s to support the PHI variant (McCall 1991) of IBIS (Kunz and Rittel 1970). In this project, fundamental issues arose about how the integration of CAD graphics and PHI rationale should work from the standpoint of human-computer interaction (HCI). These issues were ultimately settled not by working directly on PHIDIAS but by working on the JANUS system.

JANUS combined the functionality of the CRACK system (Fischer and Morch 1988) for kitchen design with hypertext functionality needed for PHI-based design rationale. CRACK enabled designers to create kitchen layouts using a *domain-oriented construction kit*. A construction kit is a collection of graphical building blocks that can be dragged and dropped into place in a CAD system. A construction kit is *domain oriented* if its building blocks represent high-level domain concepts, such as walls, windows and stoves, sinks, etc. rather than low-level computer graphics concepts such as points, lines and shapes. Domain-oriented construction kits were used because they enabled designers to rapidly and intuitively build designs. Such a construction kit is, in essence, simply a conventional CAD *symbol library* to which semantics had been added so that each type of building block indicates what type of real-world object it denotes—e.g. window, door, stove or sink.

In CRACK the semantic information of the building block is used by a critiquing system that "looks over the shoulders" of designers as they work and points out violations of rules-of-thumb for kitchen design. An example of such a critique might be, "Do not put the stove in front of a window." The rationale for this critique is that placing a stove in front of a window creates several potential problems: 1) a person might reach over the stove to open or close the window, thus creating the risk that the person might knock over a pot or lean into the flame of burner on the stove; 2) curtains on the window might catch fire; 3) the windows might get greasy; 4) someone cooking at a stove might get distracted by looking out the window. CRACK, however, did not display this rationale for users; it only displayed a brief critiquing message.

CRACK was meant as an improvement over an expert system approach in the sense that it empowered users by both providing expert advice but allowing those users to ignore this advice when they chose. The problem with CRACK was that although it presented advice, it did not present the rationale behind that advice. Users were thus often uncertain about whether to follow the advice and how to act if they chose not to follow the advice. This deficiency was remedied by creating a new system, called JANUS, that combined the CRACK functionality with hypertext functionality that displayed the rationale for each critique using PHI. The new system had two fundamentally different kinds of functionality: support for constructing designs (using construction kits) and support for design rationale. It was therefore named JANUS, after the Roman god with two faces.

JANUS presented the rationale for critiques in the form of domain-oriented issue base (DOIB) structured using PHI. These are collections of issues, positions, arguments and subissues that commonly arise in a particular design domain. DOIBs had been developed since the late 1970s for a variety of domains including design of residences, lunar and Mars habitats, neighborhood shopping areas, health care policy and information retrieval systems. The JANUS system's DOIB provided issue-based information that was relevant to a wide range of kitchen design projects.

JANUS was successful not only in further empowering its users; it also answered the crucial questions raised in the PHIDIAS project about how and when to integrate support for rationale with support for CAD. After the JANUS system was implemented and judged successful, it was realized that these successes were actually implied by Schön's theory of Reflective Practice (Schön 1983).

Schön had viewed design as consisting of a repeated alternation between two processes, that he labeled Knowing-in-Action and Reflection-in-Action. Knowing-in-Action is the process of intuitively creating the form of a design—e.g. using pencils or CAD systems. It is a non-reflective process of unselfconscious engagement in the task of forming the design. This process continues until there is a *breakdown* of intuition when something unexpected happens. In conventional design breakdowns correspond to the designer realizing the something is wrong with the design or that some unforeseen opportunity has arisen for improving the design. Once a breakdown has occurred, the designer changes to the mental process Schön calls Reflection-in-Action. This consists of reflecting on how to deal with the breakdown situation. This is a process of critical thinking in which the reasoning behind the design becomes explicit and it cannot be done simultaneously with the intuitive process of Knowing-in-Action. Once the designer has determined how to deal with the breakdown, Knowing-in-Action takes over again and implements the solution to the breakdown.

The JANUS group saw the intuitive construction of designs using construction kits as a clear example of Knowing-in-Action. Critiques corresponded to potential breakdowns. The PHI-based presentation of rationale for critiques provided support for the designers' Reflection-in-Action.

The JANUS functionality was integrated into PHIDIAS and then additional functionality was added. JANUS' hypertext functionality was implemented using the Document Examiner, which supported display of rationale but provided no support for authoring. Because PHIDIAS was based on MIKROPLIS, it also supported authoring of rationale, thus enabling designers to add their rationale to the DOIB used by the system. This authoring of rationale was accomplished using a prototyping mechanism that enabled creating a virtual copy of the DOIB. This enabled designers to add there rationale to the DOIB and even "edit" the DOIB without actually altering the original DOIB itself.

PHIDIAS also expanded the kind of knowledge-based critiquing available. In addition to critics that fired when designers positioned construction kit building blocks in the model of the designed artifact, PHIDIAS provided critics and rationale for the selection of building blocks from alternatives. PHIDIAS also provided knowledge-based agents that alerted members of design teams to potential conflicts between their work and the work of other designers in the team (McCall and Johnson 1997). PHIDIAS was applied to a variety of design domains, including layout of computer networks in buildings, design of lunar habitats and, of course, kitchen design.

4.2.2 Discussion

Critiquing is the most prominent feature of JANUS and PHIDIAS, but it is not the most important in its implications for rationale research in SE. The most important is the theory of Reflective Practice that these systems support. A central tenet of this theory is that it is a mistake to attempt to explicitly record the rationale for the process of Knowing-in-Action. This means that the traditional approach to rationale capture cannot be made to work for this part of the design process. The reason for this, according to Schön, is that forcing humans to make the reasoning behind Knowing-in-Action explicit would prevent Knowing-in-Action from taking place. But, if Knowing-in-Action cannot happen, then neither can design, at least according to Schön. The significance of this claim is that, if true, 1) it would go a long way towards explaining why it has proved so difficult to capture design rationale, and 2) it would imply that the traditional approach to the capture of design rationale can only succeed in capture *part* of the reasoning that goes into decision making in design. This does not mean, however, that capture is not possible, merely that it is not possible if one asks the person engaging in Knowing-in-Action to record the rationale. Such capture might effectively be accomplished by another person or by automated means such as those used by Myers et al. and described below.

One important contribution of critiquing is that it alerts decision makers the existence of rationale for a decision task without their having to ask whether it exists. This is a valuable contribution that makes it less likely that the decision makers will miss valuable information. But critiquing is not the only mechanism that can do this. PHIDIAS also employs other mechanisms that detect what task designers are engaged in and alert them designer to rationale for this task, thus implementing a general sort of task-based indexing of rationale in addition to its critiquing. Burge has implemented mechanisms in the Eclipse IDE that alert implementers to the existence of rationale relevant to particular pieces of code that they look at. This is somewhat similar to the task-based indexing in PHIDIAS, but there is the question of whether Burge's approach could be extended to include a more general task-based indexing for SE.

One feature of both PHIDIAS and JANUS appears to have straight-forward application to every activity of SE. That feature is the use of Domain-Oriented Issue Bases (DOIBs). Because any decision task can be represented as an issue, DOIBs would seem to be applicable to decision tasks of all types, including those for requirements determination, design, construction, testing and maintenance.

Adapting the critiquing of JANUS and PHIDIAS to SE support systems presents an interesting challenge. This sort of critiquing is heavily dependent on CAD systems use *iconic models*. Iconic models are graphical models in 2D or 3D Euclidean space of artifacts that occupy 3D Euclidean space. In iconic models there is natural correspondence—or *natural mapping*—between parts of the model and the parts of the real-world object it represents. In addition, the placement of a single element into an iconic model implies the existence of a whole battery of relationships with other elements in the model. These relationships include distance between elements, whether they are lined-up vertically, horizontally or at an angle, whether they are co-linear—and so forth. All the critics in JANUS and PHIDIAS are based on these implied relationships.

The only place that SE uses iconic models is in the design of graphical user interfaces (GUIs). This is therefore the one area where the approaches used in JANUS and PHIDIAS—as well as other systems described in this chapter—might find direct application to software projects.

Software designers generally create and use *symbolic models* rather than iconic models. By definition, the denotation relationships between elements of a symbolic model and the artifact it represents are arbitrary social conventions. Symbolic models can, however, come to feel like they also have a natural mapping if the relationships between symbols and real objects are truly standard, i.e. something universally accepted within a large group of people. The more software designers use models with standardized semantic meaning, the more natural this mapping will seem.

An open research question is whether the sorts of rich collection of implied relationships found in iconic models can also be found in symbolic models. Since these models often take the form of graphs, it may well be that graphic theory might provide a way of deducing such relationships. Perhaps conceptual schemas dealing with the types of elements and relationships amongst them could be used as the basis of critiquing in symbolic models. Whether a significant set of critics for SE can be developed remains to be demonstrated.

4.3 Automating the Capture of Design Rationale with CAD

4.3.1 The Rationale Capture Problem

This book emphasizes repeatedly that the biggest challenge facing the use of rationale in real-world projects is *the rationale capture problem*. This is the fact that it is extremely difficult to capture rationale in a real-world setting. The hallmark of this problem is that those involved in design and other SE activities often seem reluctant to record their rationale. Why this should be and what to do about it are controversial questions in current rationale research in SE as well as in other fields where rationale research is done.

Researchers in increasing numbers have come to the conclusion that the capture problem results from the intrusive and time-consuming nature of the traditional approach to rationale capture. In this approach, rationale must be structured according a given schema, such as IBIS, DRL or QOC, in order to be recorded. In other words, the initial recording of the rationale is in a structured form. There is little debate about the fact that this structuring process is labor-intensive process, but some claim that it is also disruptive of the free flow of intuitive and creativity thought in problem solving. Marshall and Shipman see all mandatory structuring as inhibiting user input (Marshall and Shipman 1999), and Fischer and his colleagues use Schön's theory of Reflective Practice to argue that the explicitly structured reflection interferes with the intuitive problem-solving process that Schön calls Knowing-in-Actions.

On the other side of the debate are those who acknowledge that the capture process is intrusive and labor intensive but argue that it is worth it because of the benefits from having captured rationale and even from the process of structuring it. In the latter case, advocates of the traditional approach claim that the structuring process helps artifact developers to improve the consistency and thoroughness of their reasoning.

4.3.2 Solution Approach: Automating the Capture of Rationale

Myers, Zumel and Garcia have done research on rationale for the design of physical artifacts (Myers et al. 1999), and they are among those who see the traditional approach as the central cause of the capture problem. Their strategy for solving this problem is to automate rationale capture to the greatest extent possible. In other words, they seek to use automated computer methods to capture rationale in a manner that is so unobtrusive that a designer can be completely unaware that capture is taking place. More specifically, they adopt the *generative paradigm* of Gruber and Russell (1996) and attempt to derive rationale from data obtained during design. Interestingly, they do not use the argument for this approach given by Gruber and Russell, which is that it is not possible during design to predict what rationale will be needed later. Instead, they use the argument that the unobtrusiveness of the approach is the decisive factor.

Myers and her collaborators adopt the approach of first recording the behavior of designers using a CAD system and secondly inferring from these records both a *design history* and *design intent*. A *design history* is an account of *what* designers did and *when* they did it; *design intent* is *why* they did what they did. The goal here is not to automate all rationale capture, but instead to automate capture of "important but low-level aspects of the design process," so that designers can limit their documentation efforts to the higher-level, "creative and unusual aspects" (Myers et al. 1999). The central insight on which their approach is based is CAD systems often enable designers to perform operations on artifacts that are semantically meaningful in the application domain.

To derive a design history, they capture records of the atomic actions possible with the CAD system and then attempt to infer designers' behavior at higher levels of abstraction (lower levels of granularity). They derive a hierarchical account of designer behavior in terms of episodes created by grouping atomic actions. They also characterize the artifact in hierarchical terms as assemblies, subassemblies and other groupings of parts. From the hierarchies of behavior and artifact structure they deduce *what decision tasks designers are undertaking and what decision alternatives they are exploring*. These decision tasks all have to do with determining features of the artifact; so they correspond to *questions* in QOC rather than the more general concept of *issues* in IBIS. The decision alternatives thus correspond to QOC *options*. It should be noted, however, that the analyses of Myers et al. make no reference to QOC or any other rationale schema.

To derive design intent, they use AI techniques that speculate on user motives using so-called *design metaphors* and a formally stated set of *requirements* for the artifact. Design Metaphors are sequences of designer activities that suggest explanations for these activities.

4.3.3 Implementation: The Rationale Construction Framework

Myers, Zumel and Garcia created a software system called the Rationale Construction Framework (RCF) to implement and test their ideas about automatic capture of design rationale. RCF has three main components:

- An enhanced CAD tool
- A Monitoring module
- A Rationale Generation module (RGM)

The CAD tool used was the commercially available MicroStation95, which had capabilities for modeling in the domain of electro-mechanical design, in which the ideas for automated rationale capture were to be tested. This tool was enhanced to enable designers to indicate the semantic type of graphical objects together with type-specific semantic attributes. For example, a given graphical object might be assigned the semantic type *gear* and given gear-specific attributes such as *number of teeth* and *gear ratio*. A second augmentation of the CAD tool added a set of analysis programs linked directly to objects in the CAD drawing. A third augmentation added the ability for designers to select graphical objects from a predefined library of semantically meaningful graphical objects.

The Monitoring module in RCF unobtrusively tracks the operations of the designer with the CAD system. Those operations that are relevant to design rationale are then passed on to the RGM in real time. Such operations include creation, deletion and modification of design objects, selection of such object from the library and their use as parts of other objects, as well as the assignment of semantic information to objects. Undoing and redoing are also passed on to the RGM as is the use of analysis programs.

The RGM performs the majority of the inference done by the RCF system. It constructs *a symbolic model of the artifact* being designed. It the uses this model and the *design event log* received from the Monitoring module together with a formally specified set of *design requirements* and the *design metaphors* to construct the design rationale.

To derive *design intent*, the RGM focuses on explaining the changes to the artifact model during design. Design metaphors play a major role in explaining these changes. Two examples of such metaphors are *refinement* and *part substitution*. Other metaphors help to identify important relationships between object that are not formally indicated in the model. Such metaphors can detect when objects are created, modified and deleted together.

Identification of relationships between design requirements and the changes in design objects also plays a crucial role in deriving design intent. Specifically, RCF constructs hypotheses that such changes are attempts to satisfy requirements. Such hypotheses can be constructed with or without domain-specific background knowledge, though the latter provides richer accounts of design intent. Once hypotheses are constructed, they can then be supported or undermined by further evidence collected as the design effort proceeds.

Myers, Zumel and Garcia (1999) describe the testing of RCF in a project aimed at designing a three-degree-of-freedom surgical robot arm. The system recorded and analyzed design activities from initial design through multiple stages of refinement. RCF was successful in describing designer activities at several levels of abstraction, identifying stages where the designer concentrated on revisions of particular parts or subassemblies, identifying the results of design tradeoffs and in explaining key changes in the design.

4.3.4 Discussion

The rationale capture problem is of such importance for the future of rationale usage that a claim to capture a significant portion of it automatically cannot be ignored. The work on the RCF looks like a promising extension of research on domain-oriented design environments. Myers et al. have used the same sort of semantically meaningful components found in the construction kits of JANUS and PHIDIAS, information used by those systems to identify design decision tasks, decision alternatives and decisions taken. But RCF's abilities to identify and characterize designer activities and to speculate on the reasons for them goes far beyond what JANUS and PHIDIAS can offer. RCF approach provides a way of capturing rationale for the intuitive Knowing-in-Action that Schön claims is disrupted by the explicit reflection that traditional rationale capture requires.

RCF, like JANUS and PHIDIAS, relies on the natural semantic mapping available in the iconic models that CAD systems create. This means that there is a question about how well the RCF approach would transfer to the purely symbolic models that are used in software design. But to the degree that the symbols used in software design models are genuine standards and not the arbitrary creations of individual designers, transfer would seem to be possible.

If transfer of the principles of RCF to software design is possible the benefits would be considerable. Of prime importance, of course, is that it might solve at least part of the capture problem. But in addition, RCF's emphasis on rationale for *explaining changes* has crucial implications for change analysis as well as the iterative and evolutionary methods of software development.

4.4 Parameter Dependency Networks as Design Rationale

4.4.1 The DRIVE System and Parameter Dependency Networks

De la Garza and Alcantara (1997) describe a software system, called DRIVE (Design Rationale in Value Engineering), that provides additional computer support to the aid of designers who document their rationale. As is often the case, the additional computer support requires a higher level of formalization of rationale than is common with most rationale management approaches. The DRIVE approach, however, can be viewed as a simple extension of the formalization required for Design Space Analysis in QOC.

The DRIVE system enables designers of physical artifacts to create dependency relationships between the parameters of objects found in a model of a physical artifact that is being designed. Such dependencies can then be used as rationale for design decisions made using a CAD (computer aided design) subsystem. More specifically, these dependencies constitute rules—or more accurately, rules of thumb—for design decisions. These rules can then be used to critique the decisions that the designer makes using CAD. The DRIVE system uses these rules to detect conflicts created by decisions and then alerts the designer to the existence of the conflicts as they use the CAD subsystem. The designer can then either resolve the conflicts immediately or postpone the resolution. Conflict resolution is accomplished by altering the design, altering the dependency rule or canceling the dependency rule for a specific CAD decision.

There are two types of parameter dependencies that DRIVE supports. One type is the dependency of the value of a parameter (attribute) of an object on the value of a parameter of an object, where either the parameters are different or the objects are different or both. The second type is a dependency of a parameter constraint on the values of other parameters. There are also two ways in which dependencies can be represented: as mathematical formula or as an if-then rule. The following is an example of an if-then dependency of a parameter constraint on a parameter value as it would be expressed in the DRIVE system (de la Garza and Alcantara 1997):

> If [Mechanical Room]:[General Function]
>   Is equal to "House Mechanical Equipment"
> Then [Mechanical Room]:[Fire Resistance Rating]
>   (minimum value) is not less than 2 hours

In ordinary language this rule says that if the general function of a "Mechanical Room" is to house mechanical equipment, then this room should have a fire resistance rating of at least 2 hours. It should be noted that in DRIVE each such rule is accompanied by natural language text that explains the rule and can provide additional arguments for them.

4.4.2 Discussion

4.4.2.1 How DRIVE's Parameter Dependency Networks Relate to Other Approaches to Rationale

DRIVE's treatment of rationale resembles QOC's Design Space Analysis in the sense that it deals only with rationale for features of the designed artifact. But DRIVE's description of artifact features is more specific than QOC's. QOC only provides a textual description of a feature, but DRIVE provides a three-part feature description: 1) a type of object, 2) a parameter (i.e. an attribute) of the object, and 3) one or more allowed values of that parameter. While QOC, like DRL, evaluates a proposed artifact feature by means of assessments with respect to criteria, DRIVE assesses a proposed decision about a parameter value by means of other parameter values.

The DRIVE system resembles both JANUS and PHIDIAS in its use of a critiquing system that delivers textual rationale to designers of physical artifacts as they work in a CAD system. The crucial innovations of DRIVE are 1) use of parameter dependency networks as the basis for critiquing and 2) enabling designers to create their own critiquing rules.

The dependency relationships used by de la Garza and Alcantara in DRIVE are more specific that the dependency relationships used by Burge in RATSpeak and her SEURAT software. Burge's dependencies are natural language arguments that do not enable the computation of values and constraints as in DRIVE. And, of course, DRIVE's parameter dependency networks are far more specific than the dependency network between decisions (issues) that PHI uses to structure rationale.

As with JANUS, PHIDIAS and the Rationale Capture Framework of Myers et al., DRIVE depends on the natural association of semantic meaning with the graphical objects used in the CAD system, i.e. that natural mapping of iconic models. This is crucial because the critiquing depends on the rules applying to classes of objects. In DRIVE, this is, in effect, accomplished using *is-a* and *has-a* relationships, though the actual implementation of these concepts is domain-dependent and complex.

4.4.2.2 Significance for Software Engineering Rationale

DRIVE's use of algebraic formulas for dependencies seems unlikely to find extensive application in SE, but its if-then dependencies would seem to have a wide range of applications in SE. They constitute a more specific and more computable version of the argumentative dependencies between decision alternatives found in RATspeak. This is especially significant in view of the fact that RATSpeak was created in an attempt to tailor DRL to the needs of software engineers who do maintenance. The if-then computational dependencies used in DRIVE are especially promising for change analysis, which is one of the most important and popular of uses of rationale in SE. Investigating the potential value of parameter dependency networks should therefore be an important topic for future research on software engineering rationale.

4.5 Case-Based Reasoning as Design Rationale

4.5.1. From Automated Case-Based Reasoning to Case-Based Design Aids

Case-Based Reasoning (CBR) (Riesbeck and Schank 1989) (Kolodner 1993) began as a branch of artificial intelligence (AI) research. It was meant as an alternative to the dominant AI approach, sometimes called Model-Based Reasoning (MBR). MBR had run into well-known difficulties, and CBR researchers thought their approach offered a way around many of these difficulties. MBR is about reasoning from principles, often in the form of *rules* or *productions*. Roughly speaking, CBR does not reason from principles but from similarity of a current problem to cases of previously solved problems.

CBR originally dealt with the creation of automated systems that mimicked the human ability to use knowledge of prior cases to deal with new kinds of problems. But it eventually became clear that the number and complexity of cases it could deal with in a completely automated manner was quite limited (Narayanan and Kolodner 1995). To address these problems, Kolodner began to look at developing non-automated CBR systems that aided human problem solvers in complex problem domains. The idea was that by learning how to aid humans who solved complex problems, CBR researchers would get better insights about how these problem solvers use large collections of complex and often incomplete cases. These insights could ultimately be used to improve fully automated CBR systems.

The primary applications domain chosen for study was architectural design, i.e. the design of buildings. Kolodner and her computer-science colleagues at Georgia Tech worked with faculty and students in the Department of Architecture at that institution to create Case Based Design Aids (CBDAs) and populate them with information about buildings. This effort resulted in a number of systems, including two versions of the ARCHIE CBDA for building design and DesignMuse, a generalized authoring tool for creating CBDAs for different domains of physical artifact design. Originally the building domain was restricted to the design of courthouses, but later it was expanded to deal with libraries and tall buildings.

CBDAs are case libraries for design, i.e. "structured, indexed and searchable databases of analyzed case studies" (Narayanan and Kolodner 1995) containing descriptions and evaluations of existing designs, e.g. the designs of existing buildings. The descriptions are typically represented using multiple media, including text and graphics. The purpose of a CBDA is to provide information about lessons learned from the experiences of previous designs so that current designers can avoid pitfalls of past projects and can benefit from solution ideas that have proved successful in such projects.

CBDAs contain cases structured around four major categories of information: *descriptions, problems, stories,* and *responses. Descriptions* are multimedia representations of designed physical artifacts. In ARCHIE these take the form of annotated CAD drawings of floor plans, elevations and sections of buildings, as well as sketches, photographs and animations.

Problems are descriptions of unresolved conflicts that are common and persistent in a type of building. The following is an example of a problem statement:

> Clerestories [narrow, horizontal bands of windows just beneath a ceiling] and skylights can help light large interior spaces, but they can also cause costly environmental problems. They can create hot spots in warm weather and increase air-conditioning costs (Zimring, Do, Domeshek, Kolodner 1995).

Stories are brief representations, in text and other media, of how the problem or a solution has manifested itself in a particular building. The following is an example of a story about a solution for the above-given problem is as follows:

> In the Gwinnett County Courthouse clerestories and skylights were used to illuminate the interior atriums. The high, angled skylights are made of tinted glass. The depth and tinting of the skylights helps prevent direct sunlight from flooding the building (Zimring, Do, Domeshek, Kolodner 1995).

There can be many stories for a given problem. The same is true for problems and responses.

Responses are general strategies a designer might consider for resolving the problems. There can be many suggested responses for each problem. The following is an example of a multi-point response to the above-stated problem:

> Use tinted glass where possible. Use clerestories rather than skylights. Angle and inset skylights to block direct sun. Use electronically moveable/controllable louvers (Zimring et al. 1995).

There can be many such responses to a problem.

CBDAs enable designers to retrieve information either by using special case-based retrieval mechanisms or by browsing using hypertext links. One of the special retrieval mechanisms automatically retrieves relevant cases based on the designer's description of a current problem's goals and constraints. The other uses an induction algorithm that clusters cases to build a hierarchical index. Hypertext links in ARCHIE and other CBDAs connect design descriptions to stories, stories to problems and problems to responses (Zimring et al. 1995).

4.5.2 Discussion

4.5.2.1 Design Case Libraries as Design Rationale

The creators of ARCHIE and other CBDAs make no claim that the information in these systems is design rationale (DR); yet there seems to be little reason to doubt that it is. After all, the cased-based information in ARCHIE deals with design problems, design solutions and solution strategies. It includes descriptions and evaluations of designed artifacts. Its sole function is to provide information that can help designers to make better decisions, i.e. to aid designers' reasoning. And, as with almost all other rationale approaches, the information in a CBDA is organized as a hyperdocument of links and nodes of text and graphics.

While case-based information about design clearly must be counted as design rationale, it differs profoundly from all other known types of design rationale hyperdocuments, including those based on IBIS, PHI, QOC, DRL, SCA or any of the SE-specific approaches currently in existence. CBDAs provide a fundamentally different perspective on how to go about collecting structuring, indexing, retrieving and using design rationale. And this new perspective comes with a solid intellectual pedigree in cognitive science and computer science. No picture of research on rationale would be complete if it omitted the work on CBDAs like ARCHIE. A crucial task for future rationale research will be to fit case-based design rationale into the overall landscape of rationale approaches.

4.5.2.2 Design Case Libraries as an Alternative Approach to Reuse of Rationale

In the rationale research literature there have been two main approaches to reuse of rationale. One is the addition of rationale to design patterns. The other is the use of domain-oriented issues bases (DOIBs). Design case libraries represent a third fundamental alternative.

One way to attempt to understand the crucial differences between the three alternative approaches to rationale reuse is to compare the ways in which they use generalization and specificity in reasoning about new projects. Rationale linked to patterns represents an attempt to create generalized stores of reasoning, in other words, collections of rationale that involve generalizations that apply to many specific design projects. This can be seen as in effect reasoning from principles, the central notion of MBR in AI research. Case libraries for design, however, are based on a fundamentally different approach to reasoning, namely CBR, which involves reasoning from cases of previous, specific projects to draw conclusions about a current, specific project.

DOIBs involves a type of reasoning that falls in between the MBR-type of reasoning of pattern-based rationale and the CBR reasoning of case-based design rationale. Where it sits in between MBR and CBR depends on which of two distinct modes a DOIB is used in. One mode attempts to create collection of texts—including issues, positions and arguments—that can be re-used *as is* in many projects. This mode is exemplified by the use of the DOIB for kitchen design in JANUS. In its reuse of unmodified information in many specific projects this mode is like pattern-based rationale except that there is no claim of either completeness or correctness for the texts in the DOIB.

A second mode of use of DOIBs is provided by the virtual copying of hypermedia networks that is available in PHIDIAS. This enables the creation of a new DOIB by making and modifying a virtual copy of the original DOIB using the prototyping inheritance mechanism in PHIDIAS. This is typically used to create a more specific DOIB than the original, in particular, one tailored to a specific project. This mode of DOIB usage is in between the general-to-specific reasoning of pattern-based rationale and the specific-to-specific reasoning of case-based design rationale, because it uses generalized information but adapts it to a particular project.

There is also a third way in which the hypermedia network inheritance functionality of PHIDIAS can be used. In this approach a new project-specific issue base is created by virtually copying and modifying a previous project-specific issue base. This approach takes a significant further step towards the type of reasoning used in cased-based design rationale, but the schema for issue-based rationale remains dramatically different from the schema for cased-based rationale of CBDAs like ARCHIE.

4.5.2.3 The Relevance of Cased-Based Design Rationale to Software Engineering

Despite the fact that CBDAs have be created only for the domain of physical artifact design, there seems to be no fundamental reason why they could not be applied to software design and perhaps even to the full spectrum of development and maintenance activities in SE. Given the fact the case-based approach to rationale is so fundamentally different from other rationale approaches, exploring its potential for SE would seem to be an important topic for research in software engineering rationale.

Where case-based design rationale would appear to have special promise is in the design of human-computer interaction (HCI), because it is fundamentally a user-centric, rather than decision-centric, approach to rationale. Currently, there is only one user-centered approach to rationale that is usable for this purpose, namely Scenario-Claims Analysis (SCA). In fact, the current heavy emphasis on both static and animated graphical representation of artifacts in CBDAs would be directly applicable to a case library of HCI designs. Such a case-based approach to interface design might be a useful complement to SCA, though it also seems possible that the two approaches might be integrated.

Of course, most of SE does not deal with the creation of an intrinsically graphical artifact as is the case with both physical artifact design and HCI design. But software design, like the design of physical artifacts, does involve use of models that have a graphical representation. Such models can be annotated and could easily have *problems, stories* and *responses* associated with them. While these models are purely symbolic in nature and do not have the *natural mapping* to the artifacts they represent that *iconic* models like floor plans have to buildings, this does not seem to constitute an insurmountable obstacle to the creation of CBDAs for SE.

4.6 Summary and Conclusions

There are fundamental issues to be resolved before much of the research on rationale in domains of physical artifact design can be applied to the design of software; but the ideas in this research are important enough that the effort to resolve these issues seems worthwhile. Above all, it is the value of this work in the areas of *rationale capture* and *change analysis* that recommends it to software engineers. It seems ironic that the work on change analysis has made such progress in physical artifact design, where there is generally much less change—especially change due to iteration and evolutionary development—than is characteristic of software design. It seems appropriate that software engineers endeavor to learn and benefit from this progress.

Finally, it is interesting to note that all of the projects described in this chapter in some way apply insights from artificial intelligence (AI) research to the support of rationale. In particular, all but one of these projects—the one based on Case-Based Reasoning—bring active computational aids to support the capture and retrieval of rationale in artifact creation. This suggests that researcher in SE should seek to answer the questions of what other ideas from AI and what other computational aids might support rationale not only in software design but in the full spectrum of SE activities.

5 Decision-Making in Software Engineering

This chapter examines human decision-making, its role in software engineering, and the role that rationale can play in the decision-making that occurs within software engineering.

5.1 Introduction

5.1.1 General

Software engineering can be conceived of as decision-making. Software designers work their way through the software development process essentially by making a series of decisions. Each of these decisions can itself be analyzed as a complex episode of problem solving. Each decision depends on a substantial amount knowledge and/or conjecture; each is highly constrained by prior decisions, and exerts substantial downstream constraint on future decisions.

5.1.1 Objectives of this Chapter

In this chapter, we first describe decision-making problems, both generally and with respect to software engineering. Focusing on the *weaknesses* of human decision-making is a traditional approach in psychology and decision science, and leads immediately to ideas about how to support and improve human decision-making – specifically to avoid those characteristic weaknesses.

We then consider decision-making in software engineering as naturalistic decision-making in the sense this term is used by Klein (1997). Klein makes the important observation that humans are actually quite accomplished decision makers, as evidenced by our successful performance in many complex and risky task circumstances. Naturalistic decision-making focuses on identifying and analyzing the *strengths* of human decision-making.

Finally, we consider rationale as a resource for and an outcome of human decision-making, specifically in the context of software engineering. Decisions that have already been made, and whose consequences can therefore be observed and assessed, are the best possible guidance we can have for future decisions.

5.2 Decision-Making Problems

5.2.1 Where Decisions Go Wrong

During the 1970s and 1980s, Kahneman, Tversky, and their colleagues conducted an impressive series of investigations into human decision-making (Kahneman and Tversky 2000). The core contribution of this work is a detailed inventory and analysis of about two-dozen characteristic biases of human decision-making. For example, the *confirmation bias* is the tendency of human decision-makers to seek and to prize data that confirms their decisions over data that disconfirms their decisions. The *familiarity bias* is the tendency of human decision-makers to consider familiar data and interpretations as typical.

It is easy to see how such biases could undermine decision-making outcomes. Consider the confirmation bias. Initial decisions are frequently based on inadequate or misleading data just because better data takes more time to identify, collate, and interpret, and accordingly becomes available later in the course of investigation. For example, so-called *emergent requirements* are often critical, but identified only after initial prototypes have been developed (Brooks 1995). If data that enter the decision process "late" are employed primarily to confirm initial decisions, poor decisions will tend to be confirmed not eliminated. In such a process, emergent requirements will rarely trigger a change of direction in system development, which is arguably their primary raison d'etre.

The familiarity bias would cause a software engineer to weigh his or her own professional experience too highly, misinterpreting what is personally familiar, but possibly idiosyncratic, as being universal. For example, a designer might justify a decision by asserting that people in general need and desire a particular feature or function, when it is only the designer that actually experiences this need or preference. At the root of the familiarity bias is the assumption on the part of decision-makers that they are more or less *just* like everyone else.

Nigel Cross (2003) comprehensively surveyed empirical studies of designers, and identified several further biases that appear to be specialized for human decision-making and problem-solving in the context of design. The strongest of these is the *solution-first bias*, the tendency of designers to rapidly frame a solution to a problem they do not yet fully understand. This initial solution is then used as a vehicle to explore the problem further (see also Lawson 1979; Carroll 2000). Interestingly, the solution-first bias is actually *more* pronounced in the strategies of more experienced designers than it is in the work activity of less experienced designers (Lloyd and Scott 1994).

It easy to deduce that the solution-first bias and the confirmation bias jointly entail a behavior pattern in which designers rapidly make solution decisions before adequately understanding the full problem space, and then disproportionately adduce confirmatory evidence to justify what was quite likely an ill-considered decision. Indeed, Cross (2003) reviews considerable evidence of this pattern, which he calls *fixation*, from many design domains, including software engineering (Guindon 1990). He also reviews evidence that decision fixation causes poor design results (Smith and Tjandra 1998).

5.2.2 Poor Decisions in Software

With respect to the characteristic weaknesses of human decision-making, there is no reason to think that software is special. Brooks' (1995) classic discussion of emergent requirements in the IBM System 360 project is a clear instance of solution-first design aggravated by the confirmation bias, and leading to poor design decisions. His famous conclusion that designers need to be prepared to "throw one away" is a strategic orientation to managing these characteristic biases.

However, decision-making is not a topic that is energetically focused upon in software engineering research. For example, Ngo-The and Ruhe (2005) surveyed the requirements engineering technical literature for the five years 2000-2005 and found only 44 articles that addressed decision-making. In general, software engineering has approached decision-making *normatively* – seeking to avoid the pitfalls of human decision-making biases by enforcing structured models for decision processes. The *classical decision model* (e.g. Janis and Mann 1979) is not so much a model of how people actually and naturally make decisions as it is a prescription for how decision makers *ought* to make decisions. Table 5.1 enumerates a typical version of this model.

Table 5.1. Classical decision model

---

1. Exhaustively survey and enumerate alternatives
2. Identify criteria and cost-benefit tradeoffs for evaluating alternatives
3. Weigh each criterion (iterate until weightings are complete and consistent)
4. Rate alternatives; for each of the top alternatives, follow entailments of contingencies and interactions with respect to linked decisions
5. Pick best alternative

At first encounter, this model seems impressively comprehensive. It presents an algorithm for making an optimal decision. Who could ask for more? Ironically then, it has come as a surprise to many research and practitioners that this model is both impossible to implement and fundamentally inadequate.

The model is impossible to implement because in any decision domain of reasonable complexity, most of the steps enumerated in Table 5.1 cannot be carried out. Thus, for most software engineering decisions, one cannot enumerate a priori the space of possible alternatives. Indeed, in many complex decision domains, discovering new alternatives – at least novel variants of known types of solution strategies – is routine, and often required. Moreover, it is often not practical to take an enumeration approach because doing so would take too long, or consume too many other resources, chiefly human effort.

The problems do not stop there. Identifying criteria and cost-benefit tradeoffs is clearly important for evaluating alternatives. But it is often not possible to identify a set of criteria that are strong enough (in the sense of measurement theory) and mutually orthogonal to guarantee that a complete and consistent weighting is possible, let alone practical.

The classical model of Table 5.1 is suitable for modeling decision-making in highly constrained circumstances, such as certain games. The algorithmic nature of the model facilitates explicit computational modeling of such decision-making, and has played an important role in decision-making agents (Chaib-Draa and Dignum 2002). However, in real-world decision-making, such as the decision-making in software engineering, the classical model leads to indeterminacy and deadlock.

5.3 Naturalistic Decision-Making

5.3.1 Background

The classical model of decision-making is not only impossible to implement, it is inadequate: It fails to identify, describe, and explain anything about some of the most important aspects of human decisions and decision-making. As we have discussed earlier, Rittel (Rittel and Webber 1973) observed that decision-maker in urban planning often became lost in a web of decisions they could not keep track of, and were overwhelmed by wicked problems that fundamentally have no optimal solution, and indeed offer only a set of variously unattractive compromises. Rational, hierarchical decomposition methods cannot solve such problems; there is no single correct decomposition, and the combinatorics of problem features overwhelm any exhaustive analysis.

Gary Klein (1998), from whom we have taken the term naturalistic decision-making, describes vividly how he embarked on a 15-year program of research on decision-making. The first thing he noticed was that everything he had expected – based largely on the classical model – was wrong. For example, real expert decision-makers do not even *try* to enumerate all alternatives, and indeed they often make decisions instantaneously, without even considering a single alternative course of action.

Although studies of decision-making biases and other decision-making problems provide an important source of guidance in understanding decision-making in real domains, such as software development, these studies are themselves biased in a peculiar way. In order to eliminate the complicating influences of domain semantics, tacit expert knowledge, and of over-learned professional practices, psychologists and decision theorists often study simplistic and contrived problems.

However, this can be seen as just another bias: After all, a contrived puzzle context *is a context*. It is a serious – and open – question whether the lessons gained from studying a puzzle context can be generalized to other contexts, such as software development. Real decisions are embedded in workflows, data gathering, conversation, and planning. They are almost always on the basis of inadequate information, unclear and sometimes contradictory goals, time and other resource pressure, and relatively high costs of failure. Decisions are typically made in dynamic circumstances; if the decision-maker hesitates, the problem has changed. Decisions are not answers to puzzles printed on a page.

A second worrisome characteristic in "classical" studies of decision-making is their focus on *bias and error*. Of course, understanding these pathologies of human decision-making are vitally important for designing instruction and support for decision makers, but they are not the whole story. Humans may be biased in characteristic ways, but they are also quite good at decision-making.

5.3.2 The Recognition-Primed Decision Model

These two lines of critique come together in Klein's (1998) *recognition-primed decision model*. In this model, expert decision-making is chiefly a matter of classification. Experts experience situations as exemplars of known prototypes. If they make such a classification, and cannot immediately reject the classification as specious, then they know what to do from past experience. If they do reject a classification, they examine the next-most-likely classification, and so on.

Klein calls the recognition-primed decision strategy "satisficing", after Simon (1957). The strategy selects the first acceptable alternative. This contrasts sharply with the 5-step classical decision model enumerated in Table 5.1. Klein's model involves a variety of sophisticated intellectual mechanisms: such as intuitions (through which the decision-maker apprehends the situation holistically as a pattern in time, and evaluates qualitative expectations about change), analogical reasoning (in which the decision-maker deliberately sees aspects of the problem situation counterfactually or metaphorically in order to reason more creatively), and mental simulation (in which the decision-maker steps through an analog mental model to assess decision consequences and trajectories). But most notably and importantly, Klein's model actually accords with observations of expert human decision-makers, such as firefighting, air traffic control, aircraft operations, obstetric medicine, software engineering, and crisis management.

Decision makers recognize current circumstances as instances of patterns they have encountered before. They build models of current situations to support further exploration through what-if? reasoning, with the objective of understanding the situation just well-enough to identify a satisfactory and actionable option. These models are very rich in the sense that they incorporate a huge amount of expert domain knowledge, but they are often quite informal. Stories and analog mental models are often used because they can incorporate a lot of the expert's knowledge, and yet still be flexible and open to further elaboration and development.

Klein reports that these models are sometimes *anti-models* in the sense that they vividly present features of the problem context or of alternative outcomes that the decision-maker wants to *avoid*. This is a highly adaptive natural strategy of decision-making. The Danish ergonomist Jens Rasmussen (1974) emphasized that that error is inevitable in tasks of any complexity, and that one of the most effective strategies for curtailing the consequences of human error is to make error as visible as possible. Anti-models are cognitive tools through which experts decision-makers regulate their own potential decision-making errors.

Klein's ideas about naturalistic decision-making are very compatible with a broader revisionist movement in contemporary social, cognitive and behavioral science that has urged greater attention to what people do in real situations, sometimes called "situated cognition" (e.g. Lave 1988). The leading idea in situated cognition is that social and material contexts are resources for human cognition and action. To take a favorite example, it is much easier to reason about lumps of butter for a cookie recipe than it is to carry out multiplication of fractions. In this view, it is simply idealistic to analyze decision-making without considering that real decisions are characteristically complex problem-solving carried out in near-real time, high uncertainty, and high downside risk simultaneously constrained by political, social, organizations, human, technological, functional, temporal, budgetary, and other resource factors. As we noted in Chapter 1, rationale is a tool for benefiting from lessons learned.

5.4 Rationale as a Resource for Decision-Making

When people make decisions, they are accountable. All decision-makers know that they may be asked why a decision was made the way it was, why an alternative was selected, or why a different alternative was *not* selected. The answer to such why-questions is the decision rationale. If we think of the software development process as a lattice of decisions, unfolding in time, then rationale is the justification for each of those decisions.

Codifying and maintaining rationales can be provide guidance in decision-making by helping to evoke reflection and self-criticism. Codified rationale can be useful subsequently by summarizing the patterns and the lessons of one's professional experience, and that of other professionals.

5.4.1 Classical Decision Making

The classical decision model (Table 5.1) begins with an exhaustive enumeration of alternative decision outcomes. Rationale provides evaluation criteria and cost-benefit tradeoffs for evaluating these alternatives, as well as guidance in weighing each of the decision criteria, and using them to rate the alternatives.

As we have already observed, complex decision-making does not, and cannot follow this model. Nevertheless, as in other intellectual endeavor, ideal models have their place. As Parnas and Clements (1986) cleverly put it, there are good reasons to "fake" a rational design process. They note that ideal descriptions are often simplified to make important concepts and relationships more salient (cf. *acceleration* in the physics of frictionless planes). They note that idealizations can provide standards for reference, rather than standards anyone would or could actually follow. In this way, idealizations can help to create and sustain a professional practice – even though they do not exemplify actual practice.

Much work on design rationale in software engineering follows this model. For example, in one of few empirical studies of rationale in software engineering, Conklin and Burgess-Yakemovic (1990) showed that an explicit Issue Based Information System (Rittel and Webber 1980) rationale – visualized with their gIBIS tool – enabled several actual software errors to be identified. However, the rationale was created through a deliberate and effortful research manipulation, not through the routine and authentic practices of software engineers. The rationale was presented with an extremely powerful (for the 1980s) graphical visualization system. And the rationale was ultimately used effectively through a quirky procedural machination: Because of a software upgrade, the rationale database had to be hand-transcribed into a new storage format, and it was during this transcription that the software errors were discovered!

Conklin and Burgess-Yakemovic did not study naturalistic software development: Their study gives no reason for us to hope that software engineers can actually be coaxed into creating or using Issue Based Information Systems, and surely they did not intend to propose that transcribing rationale databases by hand should be a routine step in making use of them! Rather, they studied a research model that in effect faked a rational process. However, the study showed that under idealized circumstances it is possible both to capture and to use rationale. This is an important contribution, and has made this study a widely used pedagogical case.

5.4.2 Naturalistic Decision Making

Naturalistic decision-making emphasizes the semantics and dynamics of real world contexts of decision-making, and the considerable domain knowledge and skill of expert decision makers. This is far more than merely the logical warrant, and the line of reasoning for decision outcomes, as it is in classical decision-making. Rationale in naturalistic decision making includes the circumstances in which the decision rationales were noticed and developed, the personal experiences, stories, professional beliefs and values of the persons who articulated the rationales, the methods they used and the specific instances of observed or conjectured system behavior and user interaction that were employed in developing the rationale.

But how could that sort of rationale be captured and used? Klein (1998) suggests that the stories about episodes of practice shared among expert decision-makers are an important transmission medium for the patterns that experts recognize so quickly in actual decision-making. Wenger (1998) describes stories as a typical vehicle in professional *communities of practice* to share results about ways of doing things. Indeed, in the late 1940s Herman Kahn had described the "accidental war" scenario – in which an isolated nuclear error precipitates all-out war, a story that guided Cold War geo-politics for 50 years (Kahn 1962).

Carroll et al. (1994) videotaped stories from members of a development team at 6-month intervals over a 2-year project. Team members generally found it enjoyable to share their accounts of how issues were identified and analyzed, how project decisions were made, and what challenges were currently being faced. Interestingly, different team members often told fundamentally inconsistent stories. A digitized database of the stories was found to be especially useful in helping to quickly orient new team members (Karat et al. 1995). Constructing this video database of informal story-based rationales was arduous, though digital media tools have improved considerably since 1992.

The more general lesson is that in order to emerge from and to effectively assist naturalistic decision-making in software engineering, rationales need to be well integrated into the social activities of learning and performance in software development. Stories are a good example of a naturalistic strategy for rationale, because narrative is so fundamentally human – as evidenced by dreams, myth, folklore, and everyday human social interaction. Other complementary strategies involve integrating the creation of and access to rationale into debuggers, bug-trackers and other software development tools and environments.

Naturalistic decision-making entails a different epistemological stance toward rationale than classical decision-making. It does not necessarily – and we would argue it definitely *should* not reject the logic-based and schematic rationales of classical decision-making, rather it encompasses these and much more with respect to the situations in which rationale and its use is embedded.

5.5 Summary and Conclusions

Design rationale is both a natural strength and weakness of human decision-makers. From the standpoint of naturalistic decision-making, humans have strong desire to understand causal dynamics of outcomes. Humans are not satisfied with *faits accomplis*; they want to know why. However, rationale – whatever else it may be – is documentation, and if software engineers and their users agree on anything, it is that most documentation is too much trouble to write or to read.

The resolution of this yin/yang of rationale is to acknowledge that rationale is an essential resource for reliable and effective human decision-making. Good decisions in any domain require support for sharing and developing best practices. A key challenge in designing rationale is to evoke the interest of decision-makers in understanding and assimilating explanations of prior decisions, while not overwhelming them with information, or information management tasks.

Part 2
Uses for Rationale

One of the major stumbling blocks in rationale research has been the fear that rationale may not be worth the cost of its capture. With the continuing emphasis on software quality and process improvement, the development community has become more aware that software development is not only about producing code and that the upfront costs of a more rigorous process result in downstream savings. The question is, how can the rationale be used and do those uses justify its cost?

The first use that comes to mind for rationale is the most simple— presenting the rationale to the software developer or maintainer when they need it. The presentation of rationale (Chapter 6) helps the developer learn about the software, the criteria that guided its development, and helps them to avoid errors in future decisions.

Rationale supports the ability to evaluate decisions (Chapter 7) to ensure that the choices made during development are the ones that best meet the needs of the system stakeholders. The importance of this evaluation is highlighted in the CMMI's inclusion of the Decision Analysis and Resolution process area (CMMI Team 2006).

Software development is a highly collaborative process since most systems are well outside the size and scope where they can be developed by only a few people. The support for collaboration provided by rationale (Chapter 8) has been demonstrated by field studies (Conklin and Burgess-Yakemovic 1991) and was highlighted in a talk on collaboration presented at the Future of SE track of the 2007 International Conference on SE (Whitehead 2007).

Unlike with hardware, software engineering rationale can be directly connected to the artifacts that it describes. These connections allow rational to support change analysis (Chapter 9) by identifying inter-decision dependencies and showing the developer how changes in decisions affect the software. This, and the other uses described in this chapter, clearly indicates that rationale has the potential to provide software development assistance that far outweighs the cost and effort required to capture it.

6 Presentation of Rationale

This chapter examines issues of presentation for software engineering rationale (SER). The substance, the content of rationale is always mediated by some presentation. The presentation could be free form, natural language text, or a formal, symbolic language; it could be printed sheets of paper, or three-dimensional displays in a virtual environment. The presentation of rationale has its own effects on the utility of rationale as an information resource in software development.

6.1 Introduction

6.1.1 General

The ingenuity and effort of creating a sound and comprehensive rationale is only worthwhile if people can use it. The use of rationale is always mediated by a *presentation* of the rationale. The presentation of a rationale can be relatively formal and symbolic, for example, using types and logic with labeled links, or it can be relatively informal, such as free text or even videotaped interview with a designer explaining his or her design. Various approaches to presenting rationale themselves have rationales. One significant advantage for Software Engineering Rationale (SER) presentation is that unlike hardware devices, the designed artifacts themselves are stored electronically. This supports the potential to attach the rationale for the artifact directly to the artifact in ways that are impossible in other design domains.

6.1.2 Objectives of this Chapter

This chapter describes the main line of development of the IBIS (for Issue-Based Information System) notation for rationales, from the early innovations of Kunz and Rittel (1970), through work on hypertext and hypermedia rendering of IBIS graphs, through to studies of the use of IBIS and IBIS-derived approaches to presenting rationale. One of the key issues that emerges from this line of research is that there is a tradeoff between the discipline and clarity that obtains from casting a design discussion into as IBIS presentation, and the inflexibility and cumbersome aspects of working with IBIS. In part, these tradeoffs led to a turn toward informal presentations of rationale in the mid-1990s and subsequently. Today, reconciling these approaches, and enhancing them through new techniques in information visualization, seems feasible, and perhaps even more necessary as the role of software developer expands to include end users.

6.2 Codifying Rationale Semi-Formally

6.2.1 The Rationale for Rationale Notations

Discussions of rationale quite appropriately tend to start with Kunz and Rittel's (1970) concept of issue-based information systems (IBIS). IBIS presents rationale as structured discourse of *arguments* that support or oppose *positions* that themselves respond to *issues*. This results in a straightforward and explicit relational decomposition of issues, positions and arguments. But IBIS quickly gets more complex: arguments can support or oppose *other arguments* as well as positions, and in particular, a given argument can support/oppose arguments that pertain to other positions on other issues. Issues have many interrelationships; one issue can *illustrate* another issue, *generalize* another issue, *resolve* another issue, etc. Thus, the hierarchy of issues, positions, and arguments is actually a network.

The key insight of IBIS can be regarded as essentially presentational: Kunz and Rittel emphasized that in planning and design problem solving the key ideas, the "solutions", were often "there" in plain view, but not always identified, weighed and valued appropriately. IBIS makes explicit how the elements of a complex problem solving process interrelate. It presents the underlying argumentation as a graph of related propositions so that planners and designers can have more precise discussions.

An IBIS presentation of the status of a design rationale makes public what issues are currently identified and how they are related. This can focus disagreements and discussions and make them more productive. For example, it a debate about what the positions are with respect to an issue is a very different debate from that of how various arguments support or oppose a set of positions. It is efficient to distinguish between these two sorts of debates, among others.

An IBIS presentation of a design process can also be a generative tool. Laying out the network of currently identified issues, positions and arguments, helps to suggest further issues that need to be raised, or further relations among issues already identified. It makes clear what positions have been identified for each issue, perhaps suggesting positions that still need to be articulated. Setting out the arguments for every position shows which positions are better supported than others, suggesting where attention can be focused to strengthen and/or eliminate some of the current positions.

Thus, an IBIS presentation both explicitly codifies the current state of a problem solving process, and poses a detailed agenda for further discussion and action. It seeks to improve the outcome of deliberative processes by highlighting divergence and even controversy. It gathers and integrates the knowledge distributed among members of a planning or design project, organizing the knowledge with respect to its relevance to the project. It makes the bases of eventual decisions more transparent and auditable.

The network presentation of rationale, first developed as IBIS, has become a standard visualization for subsequent rationale projects – even those that construe the content of rationale in ways different from IBIS. For example, (Questions-Options-Criteria) is a variant of IBIS that seeks to document a design *solution*, as opposed to the discussion *process* that led to the solution (MacLean et al. 1989). Thus, where Kunz and Rittel (1970) wanted to capture and present the actual issues, positions and alternatives as they were discussed in a design process – including parts that ultimately had no tangible impact on the final design solution, QOC seeks to present only the design argumentation that justifies the design solution. MacLean, Young and Moran (1989) saw QOC rationales as themselves a form of designed documentation for a design solution. Nevertheless, QOC rationales are typically presented in graphs that are isomorphic to IBIS graphs: design Questions (essentially, IBIS issues), the Options that address them (essentially, IBIS positions), and the Criteria for assessing Options (essentially, IBIS arguments).

6.2.2 Hypermedia Presentations of Rationale

IBIS was originally conceived of as a paper-based information technology. However, as IBIS argument networks get larger and more complex, they become very difficult to read and edit in paper: They are too large for standard-sized sheets of paper, and as they change and grow, pages become cluttered with crossing lines, erasures, and annotations. And purely paper representations are not convenient to save, and very difficult to share with remote collaborators or to adapt and reuse in subsequent projects.

IBIS has been incorporated into design *war-room* practices in which a design problem is analyzed and managed through paper and string representations pinned up to the walls of a workroom (Newman and Landay 2000; Whittaker and Schwarz 1995). Wall-sized pin-up representations are large enough to display non-trivial IBIS graphs, and, relative to paper, they are easily edited. However, rooms are expensive and cumbersome in their own ways as representational media. They can't be saved for subsequent reference or reuse, and they cannot be shared with remote collaborators.

The advent of hypertext and hypermedia in the mid-1980s provided a breakthrough in the presentation of IBIS rationales. Conklin and Begeman (1988) described gIBIS (for graphical IBIS), a browsing and editing tool for navigating and managing vast rationale networks. This tool provided many of the navigation and maintenance affordances of a wall-sized pin-up display, but rendered them accessible through a workstation user interface. It made possible saving, sharing, and reusing IBIS graphs.

Many hypermedia and hypertext tools for presenting rationale have been developed. For example, McKerlie and MacLean (1993) prototyped a hypermedia QOC rationale browser that incorporated documents, diagrams, images, and other media types directly into the nodes of a QOC graph.

6.2.3 Using Semi-Formal Rationales

Semi-formal rationales lie in the gray area between notations with known properties, and free-form expressions of rationale. Through the nearly 40 years of experience with IBIS and its descendants, there has always been a tension between beliefs that the discipline of categories and links could help to focus design thinking, and beliefs that the notation could be an awkward distraction from the substance of design thinking. Indeed, Conklin and Begeman (1988) reported both patterns among their early users.

One of the benefits of semi-formal notations is that they project a template structure onto design argumentation highlighting gaps, and thereby helping to further articulate requirements. Because gIBIS was actually implemented and used (albeit mainly in research lab software development projects), it helped to identify some of the second-order challenges for rationale browsers – challenges that could only become apparent through the real use of rationale presentation tools.

For example, Conklin and Begeman (1988) noted that the use of gIBIS helped to identify some specific problems having to do with the fact the IBIS does not represent design decisions per se. Decisions are critical events in design discussions; they resolve sets of positions on an issue, selecting one position and rejecting the others. The chosen positions are often embodied as a solution element (e.g. a specific piece of code). Conklin and Begeman (1988) observed that users had to keep track of design decisions, and their associated solution elements *outside* the gIBIS system.

Conklin and Begeman considered indicating selected positions through display highlighting, to visually distinguish them from the rejected positions. However, one deficiency of this approach is that the rationale for the *decision itself* – as distinct from the rationale for the position as a response to a given issue – cannot be represented. A more comprehensive approach, also discussed by Conklin and Begeman (1988), is to create a separate layer of meta-argumentation for discussion about nodes and groups of nodes in an IBIS graph. This approach obviously adds a great deal more complexity.

In the early 1990s, influential empirical studies of the use of semi-formal rationales presented through hypermedia browsers identified substantial cognitive and social obstacles (Buckingham Shum and Hammond 1994). Indeed, these specific studies were assimilated to a more general critique of efforts to support intellectual work directly with formal and semi-formal knowledge representations (Grudin 1994; Shipman and Marshall 1999). Recent work on semi-formal rationales presented through hypermedia browsers has focused on providing a richer vocabulary of categories and data types, and more flexible user interactions (Buckingham Shum et al. 2006.).

6.3 Codifying Rationale Informally

The tradition of rationale presentations inaugurated by IBIS focused on constrained symbolic descriptions. This was intended to benefit analysts and designers by providing a relatively precise description language as well as a discipline for using the language. However, for the most part this is more of an intention, a vision of what rationale could be, rather than an achievement tout court.

The semi-formal notations, such as the standard IBIS graphs, do not actually provide very much descriptive constraint, and to the extent they do provide constraint – as in the example of including no category for decisions, the constraint were sometimes found to be inappropriate. Nevertheless, pursuing even a programmatic interest in constrained descriptions is different eschewing such concerns. Starting in the mid-1990s, less formal approaches to rationale became more common.

Many of these less formal approaches to rationale were part of a concurrent rethinking of software design, and a turn toward less formal approaches to specifications and other software design representations (Carroll 1995; Fowler 2003). A central characteristic of these approaches was (1) a focus on *narrative*: stories of workflows and other organizational processes, scenarios of user interaction, and use cases of system interactions, and (2) a deliberate compromise of semantic precision for conceptual richness. Thus, where IBIS tried to impose (albeit programmatically) conceptual austerity on planning and design – the most "wicked" of problem types, in Rittel's famous term – these latter approaches took the more naturalistic stance of confronting the wickedness first.

Scenario-Claims Analysis (SCA) conceptualizes the rationale for interactive software systems as a collection of natural language propositions (claims) that are implicit in the usage scenarios afforded by the system (Carroll and Rosson 1992; Carroll 2000). The propositions are used to identify tradeoffs in the rationale for the system. Consider a simple scenario in which a person is trying to copy text using an information system that grays out currently inappropriate/disabled menu items. Going to the Edit menu *before* selecting the text to be copied, the person finds Copy grayed out. But after selecting the text, the Copy command is no longer grayed out, and the operation can be completed. This scenario illustrates a claim that graying out is an effective visual signal for currently inappropriate/disabled commands. This claim also helps identify potential tradeoffs, downsides of the graying out technique; for example, the user might not make the right interpretation; the grayed out command might just seem broken in the software, instead of suggesting that its argument needs to be specified.

SCA rationales are usually presented in tables, not as IBIS graphs, but in fact there is an obvious, though perhaps rough, mapping between the two: Each scenario in SCA presents an *issue*, or possibly a nexus of related issues. The design artifacts described in the scenario (such as the graying out technique) are *positions* that respond to the issue or issues. And the claim tradeoffs are *arguments* for and against these positions. Of course there are also differences: A user interaction scenario is both more complex and more narrow than an IBIS issue. For example, scenarios often present more than one issue, and generally illustrate only a single position for a given issue, not a range of possible positions responding in various ways to the issue. A similar comparison can be carried out for other scenario-based approaches such as Lewis, Reimann and Bell's (1996) problem-based evaluation approach in which a set of problem scenarios are identified, each presenting one or more issues, and then used to analytically evaluate a set of design proposals (positions) via an informal walkthrough (producing a set of arguments for the positions).

Other contemporaneous efforts at naturalistic capture and presentation of rationales explored narrative frameworks even less schematic than scenarios. Some of this work captured ethnographic design history material. For example, the Raison d-Etre project captured and presented the individual rationales and understanding of project members at specific points in time during a software development project. A dozen core members of a software product design team were recurrently interviewed during a 12-month period. The developers were individually asked about the goals and approaches of the project. A video database of about a thousand short clips was created (Carroll 2000; Carroll et al. 1994). The video clips could be browsed and retrieved using a set of tags (e.g. <project vision>).

This project showed that there is an abundance of rationale generated every day in software development. However, it also showed that there is only a partial convergence and consensus as to why decisions were taken, or even about what decision were taken. Developers were very interested to review and discuss the database of interview clips, but the most practical application of the Raison d'Etre materials was for helping new project members get better oriented to the issues that the project had faced, and to the diversity of positions that had been taken and of arguments that had been advanced for those positions.

Mackay, Ratzer and Janecek (2000) also employed video to capture and present design requirements, concepts, and rationales. Their approach focused on documenting a system in use by videotaping both expert users and novices in actual work contexts. They also videotaped design meetings in which new design proposals were described and critiqued. Finally, they used these real materials to plan and construct animated storyboard scenarios showing how particular design proposals might be implemented and how they might change the system in use.

6.4 Directions

The original challenge in articulating presenting rationales was the complexity and vastness of the considerations that can bear on wicked problems of planning and design. The IBIS notation brought an order to research on this challenge, but the challenge remains. Today, software technology advances in databases, and more generally in information repositories, and in data visualization present new opportunities for developments in managing rationale.

6.4.1 Reusable Rationale Databases

Since the early 1990s, papers on design rationale have suggested the possibility of repositories or libraries of rationale. Indeed, one argument that could be made for semi-formal design rationale notations is that they provide a rubric for structuring and retrieving rationale elements in such repositories. Such repositories could improve the cost-benefit balance for developing rationales in three distinct ways: They amortize the costs of developing comprehensive design rationales by permitting many authors to contribute rationale. They could improve the validity and applicability of rationales by moving the level of design discourse beyond single projects and into the entire software design community. And they could increase the benefits of developing rationales by allowing many developers to access and use rationales once they are created.

Sutcliffe and Carroll (1999) defined a structural schema for claims to facilitate claim retrieval and reuse. Their schema includes a series of labeled slots for each claim including parent claims, projected usage scenarios, design effects, upsides, downsides, issues, dependencies, evaluation data, and basis in theory. Developers could search or browse a claims repository using the values of these slots. Chewar, Bachetti, McCrickard and Booker (2005) adopted this proposal and developed a rationale repository to support the design of notification systems (interactive interface displays like RSS clients that run in background of a primary task and notify users of updates). Their LINK-UP (Leveraging Integrated Notification Knowledge with Usability Parameters) system presents claims for typical notification system scenarios. On-going evaluation of the use of LINK-UP by novice designers has been encouraging (see also Fabian et al. 2006; Payne et al. 2003).

The SEURAT (Software Engineering Using RATionale) system (Burge and Brown 2004) uses the RATSpeak representation (Burge and Brown 2003) implemented as a re-usable rationale database schema. When rationale is required for a new project, the initial rationale-base is populated with the required schema tables and a fully-populated Argument Ontology that contains a hierarchy of reasons for making software decisions. SEURAT has only been used as a single user system. The relational database would make it straightforward for multiple users to contribute rationale but there are other SEURAT capabilities, such as the ability to associate that rationale with the code, that can not be distributed using the current implementation.

6.4.2 Multi-Scale Presentations of Rationale

All of the standard presentations of rationale articulate a great amount of structure at basically a single level. This is obvious in the vast networks that gIBIS tried to manage through hypermedia browsing. However, in some ways this does not reflect the structure of a rationale space as designers and users experience it. Some issues, positions and arguments are first-order elements of the design argument; others are subordinate. But these relations are not necessarily clear or even codified at all in standard IBIS graphs.

This could be seen as an example of multi-scale data structures. For example, in a map of the world the continents and oceans are always visible, but the Hudson River may or may not be visible at that scale. However, in a map of the state of New York, the Hudson River is always visible, but the individual streets in the town of Ossining (located on the river) would most likely not be visible, though they would be in a map of Ossining or of Westchester County. The point is that map data is understood to be multi-scale data, and is typically presented in multi-scale presentations.

Analogously, rationale data might be organized so that the coarsest scale would present only the leading issues, positions and arguments. However, one could drill down to finer scales to see the subordinate issues, positions and arguments. The multi-scale concept is most typically discussed with respect to visualization techniques – as illustrated by maps. Perhaps its application to presentations of rationale should be pursued especially with respect to visualizations of design argumentation (e.g. Kirschner et al. 2003).

Wahid, Smith, Berry, Chewar and McCrickard (2004) describe a simple but concrete example from their claims repository work: They visualize a central claim as surrounded by concentric orbits of supporting or otherwise related claims. The user can filter the visualization to see only the core claim, or to see only the core claim with its most-related claims, or to see the maximum map of related claims.

6.4.3 Integrated Presentation

As stated earlier, the capture and use of rationale for software development has a significant advantage over rationale for other domains. Since software is stored entirely electronically, the rationale can be attached directly to the artifacts that it describes. This is aided significantly by progress in software development environments that have emphasized the ability to integrate and extend the various tools used in developing software. These tools include word processors used to write and access documentation, UML editing tools used in design, and the Interactive Development Environments used to write, edit, compile, and debug the code. The extensibility of software development environments has also benefited from the increasing availability and use of open source applications in these environments that provide even more flexibility and openness in customizing the environment to support and accommodate rationale.

One of the issues in the capture and use of rationale is the need for developers to record and use their rationale as a part of their normal development process. The need to have to use a separate tool for rationale has been a deterrent towards doing this. When examining past scenarios where rationale could have been beneficial in saving time or money, one question arises—would the person who could have benefited from the rationale have actually looked at it? Would they have even known that it existed? While rationale does have some benefit as a generative tool, it should not be treated as "write only" documentation.

Software design is often documented using the Unified Modeling Language (UML). Zhu and Gorton (2007) developed a UML profile that models design decisions in the UML and captures the relationships (support, break, help, hurt) between the decisions and non-functional requirements (NFRs). UML stereotypes were used to model each of these elements. The design decision stereotype describes the decision, design rules applying to the system components, design constraints, the set of architectural elements (such as UML classes) the decision refers to, and the rationale (in an unspecified format). The NFR stereotype gives attributes specific to that NFR, and the relationship stereotype describes any constraints that apply to that relationship. The profile supports consistency checking between design decisions and related architectural elements.

When building a Rationale Management System, one issue that must be addressed is how and when the developer should be informed that there is rationale available. Systems working in domains that are more constrained than software, such as the JANUS system (Fischer et al. 1989) which supported kitchen design, served as critics that presented rationale when the designers' actions appeared to contradict rules embedded in the system. The user is informed of the presence of rationale when they make a decision that appears to be incorrect. Rationale is also used interactively within a design environment in the REMAP system (Ramesh and Dhar 1992) where the rationale behind the functional specification is used to help make design choices.

While rationale can be used prescriptively to assist with designing, it is also valuable when used descriptively by providing insight into why the system is implemented the way it is. The user is more likely to be aware of, and read, the rationale behind the code if the rationale is integrated into either directly into the code that they are modifying or the environment that they are modifying it with. The SEURAT system (Burge and Brown 2004) integrated rationale capture and presentation into the Eclipse (www.eclipse.com) development framework. The rationale argumentation structure was displayed in a tree format within an Eclipse "view." In addition, three standard Eclipse views were extended/used to show the presence of rationale: the Java Package Explorer was augmented by an icon overlay on every file that had associated rationale, rationale associations were stored as Eclipse "bookmarks" and each bookmark giving an association was show in the editor used to modify code. The bookmarks could be used to jump directly from the rationale alternative to the code that implemented it. The goal behind the integration was to reduce the likelihood of a developer or maintainer working with code while oblivious to the presence of the rationale that could assist them.

6.5 Summary and Conclusions

Probably the two greatest innovations in presenting rationale are still the original information schema of IBIS and the gIBIS hypermedia browser for IBIS graphs. Some of the dichotomies that have structured research and development on rationale presentations during the past several decades have dissolved. For example, the distinction between semi-formal notations and informal notations seemed paradigmatic in the early 1990s, but probably will matter less as information systems increasingly create structure out of content, and thus do not need to force structural constraints on the humans that use them. Thus, the presentation of rationale – how it appears to its human users – will tend to matter more in the future.

7 Evaluation

Software Engineering Rationale (SER) can play several roles in supporting system evaluation. One is to support the evaluation of decision alternatives by providing the means to capture the arguments for and against each alternative. The rationale can be used to automatically calculate support for alternatives and present it to the developer to assist them in making, or revising, their decisions. Rationale also supports usability evaluation by providing a process for analyzing use scenarios via Scenario-Claims Analysis (SCA) (Carroll and Rosson 1992; Carroll 2002). In this chapter, we discuss a number of approaches for using rationale to evaluate the alternatives to assist with decision-making and also how SCA supports usability evaluation.

7.1 Introduction

7.1.1 Argumentation-Based Rationale

7.1.1.1 Decision Making in SE

Developing a software system requires making many different types of decisions. Decision-making consists of generating alternative solutions, or approaches, identifying the reasons for and against these alternatives with respect to evaluation criteria, and selecting the "best" alternative based on these reasons and criteria.

Decisions made during software development affect many aspects of the development process and the developed product:

- *Product decisions* – what is being developed? Who should it be marketed to? Who is the customer/user? What are the requirements? Where does the system need to run?

- *Process decisions* – how should the system be developed? What process model should be followed? When should versions be released? What level of documentation needs to be produced? What is the testing strategy?
- *Management decisions* – how should the development team be structured? Who should be on it? What resources should be made available to the project?
- *Development decisions* – what development tools should be used? What components can be integrated? What is the system architecture? What are the data structures?

These are only a few examples of the many different decisions and decision types that need to be made. The results of each decision may be important to a different collection of stakeholders. For example, a system user would be interested in decisions regarding functionality but not as concerned with process models or data structures.

Each decision also has several different types of criteria that influence alternative selection. These criteria include functional requirements, non-functional requirements, assumptions, dependencies, risk, and constraints. The degree to which an alternative meets or fails to meet criteria may vary as well as the certainty in that evaluation. The decision-making task is further complicated by criteria differing in importance.

7.1.1.2 Rationale and Decision Support

The information generated and used during decision-making consists of decisions required, alternatives considered, reasons for and against the alternatives, and the criteria used for evaluation. This information forms the rationale for the choices made as a software system is developed and maintained. The rationale can be used to evaluate these choices and support the human decision-maker by advising them if their decisions are inconsistent with the rationale that they recorded.

The rationale can both be evaluated itself and used support evaluation of the decisions made. Evaluating the rationale itself involves syntactic checks on the structure of the rationale and semantic checks that analyze its content (Conklin and Burgess-Yakemovic 1996). An example of a syntactic check would be to look for missing information, such as decisions where alternatives were not chosen, while semantic checks would look for contradictions in reasoning, such as arguments that are used to both support and refute an alternative.

Evaluating the decisions made involves using the rationale to indicate which alternatives are preferable over other alternatives and why. The method of evaluation and the inputs to each method vary depending on the complexity of the problem and the types of information available. Decisions may involve looking at different types of criteria (functional and non-functional requirements; assumptions; constraints, etc.), conflicting opinions from multiple decision-makers, uncertainty, shifting priorities, and missing or incomplete data. The evaluation of an alternative may change over time as well so there also needs to be a way to determine when re-evaluation is necessary.

Selecting an evaluation method requires tradeoffs between the amount of information required to use a method, the computational requirements (if evaluation is computer-assisted), and the required rigor. The value of the evaluation is directly dependent on the ability to capture the rationale in sufficient detail to support the method chosen. This chapter will describe several alternative methods for computer-assisted evaluation of argumentation-based rationale in order to augment human decision making.

7.1.2 Scenario-Based Rationale

Scenario-based design (Carroll and Rosson 1992) uses scenarios as the starting point of design. Scenarios describe how the user goes about performing a task using the artefact that is being designed. Scenarios are valuable because they are a way to take knowledge about system use that is tacit, such as assumptions, and make it concrete (Carroll 2000). Scenario-Claims Analysis (SCA) is the process of analyzing scenarios to extract "claims"—implicit causal relations that describe the desirable and undesirable consequences of design features described in the scenario (Carroll 2000). These claims describe the rationale behind the scenario—why the scenario operates the way that it does. Later in this chapter we will describe how SCA can be used in evaluation.

7.1.3 Objectives of this Chapter

This chapter discusses evaluation of and using argumentation rationale as well as using rationale generated during scenarios claims analysis for system evaluation. For the argumentation evaluation, this chapter looks at two types of evaluation: evaluation of the rationale itself for completeness and correctness and using the rationale to evaluate decision alternatives. For alternative evaluation, it concentrates on three issues: comparing the alternatives, combining inputs from multiple developers, and handling uncertainty. The focus is primarily on computational evaluation using argumentation. The scenarios claims analysis section describes how analyzing scenarios to extract claims is a form of evaluation that can be fed into development of testing scenarios to gather evaluation data.

7.2 Evaluating the Rationale

Many rationale representations take the form of semi-formal argumentation. This format is a natural way to express the decisions, alternatives, and arguments and can be read easily by people and interpreted by computer. There are many argumentation formats which date back to Toulmin's warrants, claims, datums, backings, and rebuttals (Toulmin 1958). These include the Issue Based Information System (IBIS) notation (Kunz and Rittel 1970), Questions Options and Criteria (QOC) (MacLean et al. 1989), Decision Representation Language (DRL) (Lee 1991), WinWin (Boehm and Ross 1989), the Design Recommendation and Intent Model (DRIM) (Peña-Mora et al. 1995), and numerous notations that extend these representations and Rationale Management Systems that use them.

In this section we describe two types of evaluation of the rationale: check the rationale for completeness and checking the rationale for correctness.

7.2.1 Completeness

Completeness checking over the rationale looks primarily at the syntax checks, or what Conklin and Burgess-Yakemovic referred to as "well-formedness checks" on the syntax and structure (Conklin and Burgess-Yakemovic 1995). Completeness checking typically does not ensure that all the rationale for the system has been collected but instead checks to see if there are any holes in the rationale that is present.

There are many possible checks, or inference, that can be performed on the rationale. The availability of these checks depends on the richness of the representation format. There are some checks, however, that can be made over most argumentation-based formats. These include: checks to ensure that there are alternatives proposed for each issue/decision, checks to see if an alternative has been selected for each issue/decision, checks to see if alternatives are selected that do not have any arguments (in either direction), and checks to see if alternatives are selected that only have arguments objecting to them with none in support.

7.2.2 Correctness

While syntactic inference looks at the structure of the rationale, semantic inference looks that the contents. The ability to do this is limited—comparing information within the rationale requires that a common vocabulary be used. The KBDS system (Bañares-Alcántara et al. 1995; King and Bañares-Alcántara 1997), which extends IBIS, used keywords to check argument consistency. InfoRAT (Burge and Brown 2000) created a common vocabulary of arguments. SEURAT's RATSpeak (Burge 2005), an extension of DRL, extended this vocabulary into an argument ontology that described a hierarchy of reasons for making software decisions at different levels of abstraction. Using a common vocabulary within arguments allows for inferences that look for contradictions such as using the same argument for and against an alternative.

Some rationale representations, such as RATSpeak, capture dependencies between alternatives. These relationships can be used to check if there is a dependency violation where an alternative is chosen that conflicts with another selected alternative or requires an alternative that has not been selected. If the requirements are explicitly captured in the rationale, the rationale can also be used to detect if an alternative has been selected that has an argument indicating that it violates a requirement. Some representations, such as RATSpeak and REMAP (Ramesh and Dhar 1992) represent requirements as explicit types of rationale entities. QOC and DRL can do this less directly by having QOC's critieria and DRL's goals contain requirements.

Another type of semantic inference is to detect if there have been any tradeoff violations. Many arguments captured in rationale describe qualities that are "traded off" when making decisions. Known tradeoffs that apply at a system-wide level can be captured as "background knowledge" in InfoRat (Burge and Brown 2000) and SEURAT (Burge and Brown 2004). An example of a software tradeoff would be the ease of coding an alternative versus its flexibility. In most cases, the more flexible design is likely to be more difficult to initially implement. The rationale can be evaluated to check to see if there were alternatives with arguments that claim flexibility where there were no opposing arguments warning of the potentially longer development time. The rationale can also be checked to ensure that alternatives do not claim to be flexible and easy to implement. The developer can override the results of these inferences in cases where there are exceptions to the general rule.

7.3 Evaluating the Decisions

Software development decisions are often multidimensional – decision outcomes involve multiple dimensions. Vetschera (2006) states four contributors to multidimensionality: alternatives impact multiple criteria, uncertainty of alternative outcomes, multiple stakeholders, and alternative outcomes that vary over time. The rationale can serve as inputs to many different evaluation methods. In this section we will describe some of the methods and issues and how rationale has been, or can be, used to support them.

7.3.1 Comparing Alternatives

There are many possible methods that can be used to compare alternatives. The choice of method depends on the information available as input (i.e. the richness of the rationale representation and the fidelity of the data) and results of tradeoffs between computational complexity and semantic justification of the results. Methods require extensive calculation, evaluations for each criteria, multiple pair-wise comparisons (which do not scale well if the number of alternatives is large), or quantitative measurements (which may not be available).

The simplest evaluation involves arguments that are either for or against an alternative. The support for the alternative consists of the difference in pro argument and con arguments divided by the total number of arguments (Fox and Das 2000). This method assumes that all arguments are equally important.

Many evaluation methods fall into the category of Additive Sum Methods (Vetschera 2006) where the alternative utility is calculated using a weighted value for each argument. The simplest form, Weighted Sum Method (WSM), is used by several rationale-based systems including HERMES (Karacapilidis and Papadias 2001), InfoRat (Burge and Brown 2000), and SEURAT (Burge and Brown 2004; Burge and Brown 2006). In these systems, each argument is given a weight to indicate its relative importance. Assigning these importance values is not a simple task—the values could be given relative to the specific decision or could apply system-wide. In HERMES, the evaluation involves the sum of the weights in favour minus the sum of the weights against. In InfoRat and SEURAT, the weight is applied to (multiplied by) a numerical amount indicating the degree to which the alternative affects the criteria. Additive Sum Methods can be evaluated for sensitivity to any of the weight values by plotting the result when expressed as a function of that weight (Vetschera 2006). Determining the appropriate weights can be difficult and the results of the summations do not always accurately reflect the utility. Vetschera (2006) demonstrates that a summation of weights may result in avoiding compromise alternatives. He suggests correcting this by adding an additional partial utility function to each argument in addition to the weight. This would be especially valuable when different types of arguments are involved. A violation to a functional requirement, for example, should have a significantly higher impact on the decision than other types of arguments.

The Analytic Hierarchy Process (AHP) (Saaty 1990) is another method for comparing alternatives. In this method, pair-wise comparisons are performed between all alternatives examined against all relative criteria. As with the other weighted methods, criteria are given different weights. AHP has been applied to software engineering decision problems such as prioritizing software requirements (Karlsson and Ryan 1997) and choosing software products (Lai et al. 2002). This method requires that the same criteria be used to weigh each alternative. The significant disadvantage to this method is that it does not scale well when comparing large numbers of alternatives.

7.3.2 Combining Inputs from Multiple Developers

Rationale can be a valuable tool for collaboration and negotiation. This was demonstrated with gIBIS (Conklin and Burgess-Yakemovic 1995), Compendium (Buckingham Shum 2006), and SHARED-DRIM (Peña-Mora et al. 1995). The argumentation can serve as a natural medium for the different contributors, or stakeholders, in a project to state their views on alternatives under consideration. This does pose an interesting challenge for evaluation: how can conflicting beliefs and opinions be aggregated? Factors that contribute to the difficulty include the differing expertise of developers and differing degrees of confidence in evaluations. There could potentially be arguments refuting and supporting other arguments as developers debate each other's arguments. The developers may not disagree with the arguments themselves but may not agree with information such as the importance of the argument criteria, the degree to which the alternative meets the criteria, or the plausibility of the argument.

Combining conflicting beliefs has been an important topic of research in economics, statistics, and artificial intelligence. How can conflicting beliefs be combined to reach some version of Pareto Optimality? There are numerous impossibility theories (Arrow 1963; Mongin 1999; Blackorby et al. 2000) but also many approaches that avoid impossibility by methods that include restricting the Pareto condition (Gilboa et al. 2001) and understanding that not all expert opinions should carry the same weight (Maynard-Zhang and Lehman 2003).

As with other evaluation methods, the belief combination method used will depend on the type of information available and the amount of computation that needs to be performed.

The field of economics has studied this issue when looking at preference aggregation (Andreka et al. 2002; Hild et al. 1998; Harsanyi 1955). Lexiographic ordering is another method used to combine preference operations (Andreka et al. 2002). Clemen and Winkler (1999) describe many different methods for combining probability distributions from multiple experts when performing risk analysis/assessment. These methods include the linear opinion pool (Stone 1961), which uses a weighted sum incorporating the "quality" of each expert and Bayesian updating (Winkler 1968). In AI, combining beliefs is necessary when performing ensemble learning (Pennock et al. 2000) and when merging information from multiple data sources (Booth 2002; Meyer et al. 2001).

The most promising methods are those that take advantage of information about the experts – their level of expertise, their experience, their reliability, and potentially even their influence. When experts disagree and their negotiation is captured in the rationale, they are unlikely to be given equal weight in the decision-making process and it is important to utilize this information when proposing decisions. Knowledge about the expert providing the information can be used to provide a "pedigree" for the information. This pedigree information is used in belief fusion (Maynard-Ried II and Shoham 2001) to combine beliefs from different experts.

7.3.3 Handling Uncertainty

Software decision-making needs to address the uncertainty surrounding the development process. Uncertainty can refer to many things: vagueness, imprecision, inconsistency, incompleteness, or ambiguity (Parsons 2001). Ziv et al. (1996) describe four domains where uncertainty is an issue: requirements analysis, transitioning from requirements to design and code, uncertainty in re-engineering, and uncertainty in re-use. This uncertainty can come from many sources. Three examples are the problem domain ("real world"), the solution domain, and the humans participating in the development process (Ziv et al. 1996). Lehman and Fernández-Ramil (2006) are concerned with the impact of assumptions which may change over time. When assumptions that were the basis of software decisions no longer hold they can result in system failure. A high profile example of this is the loss of the Ariane 5 rocket (Nuseibeh 1997; Lehman and Fernández-Ramil 2006). Decisions must also be made in the presence of incomplete information and may require revisitation later in the process where more is known about the problem.

The presence and role of uncertainty in making software decisions can be captured in the rationale. Systems such as REMAP (Ramaesh and Dhar 1994) and SEURAT (Burge and Brown 2006) explicitly represent assumptions in the rationale. SEURAT supports the ability to disable an assumption and re-evaluate the support level for any alternatives referring to it. If the assumption refers to an event that is expected to be true at some point in time, it should be given a time stamp to remind the designer that the decision should be re-examined (Burge et al. 2006).

The need to gather additional information can be captured in the form of questions as is done in DRL/SIBYL (Lee and Lai 1995) and SEURAT. These systems use questions to describe what information is required to make a decision or evaluate an argument and to indicate, if known, the likely sources of that information. SEURAT will report all unanswered questions as errors until they are resolved.

Uncertainty in arguments is captured in DRL, SEURAT, and Knowledge-Based Decision System (KBDS) (King and Beñares-Alcántara 1997) using plausibility, or uncertainty, values for each position. SEURAT and KBDS use these values, along with weights applied to each criteria, to rank the alternatives.

Using a plausibility value as a weighting factor in a weighted sum evaluation is one approach to incorporating the effect of uncertainty in evaluation. There are numerous other approaches that can also be used. Parsons and Hunter (1998) divide formalisms for uncertainty handling into two "camps"—the "numerical camp" that uses quantitative methods and the "symbolic camp" that uses logical, or qualitative, methods.

Numerical, or quantitative, measures include those based on probability theory, evidence theory, such as Dempster-Shaefer (Shafer 1976), and possibility theory (Zadeh 1978), based on fuzzy sets (Zadeh 1965). These methods share several drawbacks: the potential difficulty in obtaining the "numbers" (probabilities, possibilities, and distributions), the risk of comparing different types of beliefs, and the possibly significant computational expense (Parsons and Hunter 1998).

Two quantitative methods frequently used in decision-making are influence diagrams and decision trees (Clemen and Reilly 2001). Influence diagrams capture the decision structure as decisions, change events, the desired outcome (payoff node) and intermediate consequences/calculation nodes. The different alternatives, outcomes, and consequences are present as tables within the nodes. Decision-trees express this information more explicitly in the structure where decisions branch to choices and "chance events" branch to outcomes. Decision-trees are often used to compute the "Expected Value" of a decision. Decision-trees have been used to support Value Based Software Engineering by calculating the value of a software project (Erdogmus et al. 2006).

Qualitative methods are those that work either without numeric information or with only some numeric information (Parsons 2001). In some cases, these methods are variants on quantitative methods. Qualitative Probabalistic Networks (Wellman 1990; Parsons 2001) which are a variant on influence diagrams where the influence of one node on another is expressed qualitatively as being positive or negative.

Defeasible reasoning is a form or reasoning that accounts for the need to retract initial conclusions when new information is obtained (Parsons 2001). Parsons describes three forms of defeasible reasoning: logic, probability, and argumentation. Argumentation can support reasoning under uncertainty either by calculating the "safety" of arguments based on the presence of counter arguments or by adding a confidence factor indicating the degree to which the argument is believed to be true (Parsons and Hunter 1998).

The ability to re-evaluate beliefs (in our case, in the form of alternative evaluations) in the face of changing assumptions is similar to work done using Truth Maintenance Systems (TMSs) (Doyle 1979; Johan de Kleer 1986). In rationale-based systems, changing assumptions and NFR priorities can be used to re-evaluate alternatives to indicate where changes might be advisable. This process would probably stop short of actually retracting the selection of alternatives but would instead inform the developer of the potential problems.

7.4 Scenario-Based Evaluation

As described earlier, scenario-based design uses interaction scenarios as an informal and holistic working representation in requirements analysis and design. The scenarios depict user interactions observed, predicted, and proscribed, and provide a medium for exploring first-order consequences and interactions of envisioned design features. For example, one obstacle to code reuse is that it is often difficult for programmers to find examples of how a given object or module is to be reused; thus, they must work directly from code definitions, which is a strong deterrent to reuse (Rosson and Carroll 1996). In designing support for code reuse, one might envision and analyze a scenario in which part of the documentation for software objects and modules is pointers to commented example uses of that code.

The scenario might be the starting point for a design solution (e.g. part of programming environment), but it also helps to evoke and evaluate rationale. For every design feature in an envisioned scenario, one can identify desirable and undesirable consequences. Thus, providing example-based usage documentation indeed is a resource to programmers: They quickly learn to borrow usage protocol directly from example uses (Rosson and Carroll 1996). This is an upside consequence of the design solution. However, there are also downsides, risks, or costs entailed by the design solution: Positing new documentation raises the question of who will create and maintain the documentation, and of how and where it will be stored and accessed.

Evaluating a design solution and its rationale by analyzing interaction scenarios is an example of what Scriven (1967) called *intrinsic evaluation.* Intrinsic evaluation assesses solution properties analytically, instead of empirically measuring performance characteristics. Intrinsic evaluation is often more illuminating than empirical evaluation, since it constructs an arbitrarily rich decision space of implicit tradeoffs. Intrinsic evaluation can also be less expensive, But it is always less definitive in that it cannot determine the exact cost parameters in the tradeoffs. In the example of reuse documentation, the analysis identified valid desirable and undesirable consequences of the design solution, but only a large-scale implementation could show whether the benefits out-weigh the costs.

7.5 Summary and Conclusions

Here we described two ways that SER can be used to support software evaluation: supporting decision-making by evaluating decision alternatives and supporting usability evaluation through scenarios claims analysis. There are many different types of decisions made during software development that rationale can be captured for. This rationale can then be used to evaluate these decisions to ensure that choices made do not contain flaws that can be detected via computation. This evaluation is not necessarily used to make the final decision but can be used as a verification step. Evaluation is also an important aspect to change analysis by providing a means for accessing the impact of changing criteria on the recommended decisions. Scenarios and SCA evaluate how the system supports its goals in operation by providing a framework for evaluating usability based on the scenarios and the accompanying usability rationale.

8 Support for Collaboration

This chapter examines collaboration with respect to design rationale. On the one hand, this is a discussion of how collaboration can support the development, codification and use of design rationale. On the other hand, it is a discussion of how rationale supports collaboration in design and development

8.1 Introduction

8.1.1 General

People work together in software design and development because they wish to accomplish projects that are too large and complex for a single person. Although this is the fundamental basis for collaboration in all human endeavors, it is not always a simple matter of adding team members to tackle ever-greater challenges. Indeed, one of the classics of software engineering, Brooks' *Mythical Man-Month* (1975) took its title from the mistaken notion that software team productivity scales linearly with the number of team members. Brooks analyzed his own experience managing the development of the IBM Operating System 360 software, a project in which he concluded that the addition of team members eventually *reduced* productivity.

8.1.2 Objectives of this Chapter

This chapter surveys the relationship between collaboration and design rationale. First, it observed that software development is almost always collaborative work, for the simple reason that most software projects are too big for solitary individuals to ever successfully tackle. This raises a set of specific challenges: Collaboration aggregates individual efforts, but it also creates new sources of work for people in teams, and new risks for the products of team work. We then consider how collaboration supports rationale in software development – by encouraging team members to explicitly articulate their goals and plans, and therefore to create the possibility of a discussion about reasons, and by supporting a culture of software development to conventionalize and leverage social mechanisms like anthropomorphic metaphors and software patterns. Finally, we consider how rationale supports collaboration in software development – by supporting awareness of how the project is meaningful to one's collaborators, and coordination among collaborators, especially with respect to making progress in uncertainty.

8.2 Software Development as Collaborative Work

8.2.1 Collaboration is Inescapable

The most basic driver for collaborative work is the human ambition to tackle large and complex projects: Brooks estimated that Operating System 360 took 5000 person-years. Quite simply, there is just too much work to do in many projects for one person to ever be able to carry them out. But the issue is more than one of mere additions.

Collaboration is well integrated into human psychology and sociology. For example, groups of people generate more ideas and higher-quality ideas than disaggregated individuals. People with different skills and experience often experience *synergies* in collaboration; that is, together people can develop solutions that no one of them could have conceived of or executed individually (Kelly and Littman 2001).

During its brief history, software engineering has developed as a pervasively collaborative work practice. Developing a substantial software system requires many specialized skills. The tasks of system development – requirements identification and analysis, architectural specification, software design, implementation, testing – involve a great diversity of skills. Individual software professionals cannot be expert in all or even most of these skills. Indeed, software professionals typically devote a significant fraction of their professional effort to keeping up to date on just one or two of these professional skill sets.

The tasks of software development are at least partially decomposable, as suggested – perhaps a bit optimistically – by traditional waterfall models of system development. Thus, modern software development regularly involves divisions of labor and coordination of specialized contributions.

This entails fairly elaborate and articulated specializations in software project management, in addition to the primary skills of software development.

Furthermore, labor economics and the world-wide distribution of skills have produced a global-distributed paradigm for software development. Today, many systems are developed by collections of technical teams scattered throughout the world, each providing some specific capabilities, and sometimes having little or no direct contact other teams. Such far-flung projects were unprecedented only a few years ago, and still constitute an area of intense innovation in collaborative work.

In this context, the example of Operating System 360 begins to appear an unrealistically *simple* case: The OS 360 software only had to run on one hardware configuration, and was developed by a co-located team; most of the designers and developers worked in direct physical proximity.

8.2.2 Collaboration Entrains Challenges

The notion of a man-month – or person-month – is mythical in the sense that adding people to a project does not enhance to the total effort linearly. The basic reason for this is that *collaboration itself is work.* Two people chopping down a tree must share their plans and coordinate their efforts just to survive, let alone to experience a productivity boost. This sharing and coordination diverts and subtracts time and effort from the primary task. Thus, the tree may be cut down faster than either person alone could do it, but it is never cut down twice as fast.

The challenge of collaborative work is considerably greater than suggested by the tree-cutting example. When people work in groups, they tend to work less hard than they do when working as individuals – a phenomenon called "social loafing" (Karau and Williams 1993). Social loafing is especially prevalent when people perceive that their contribution to a collective outcome is not unique, that someone else could do the work just as well, or when they believe that their loafing will not be evident to their co-workers.

When people pool their ideas, when they collectively brainstorm and develop new ideas, they tend to adjust their contributions toward positions taken by others they perceive to be competent or powerful, or toward existing majority opinions – the status quo. This tendency to conform undermines the extent to which collaborative intellectual activity can generate more and better ideas, and over time causes groups to become more homogenous and less effective (Latane and Bourgeois 2001).

But diversity in groups also entails collaborative challenges. People with different technical backgrounds commonly have different fundamental values and beliefs; they can find it difficult to appreciate one another's contributions, or even understand what is being contributed (Pelled et al. 1999). Thus, diversity in collaborative groups frequently leads to conflicts, often very deep value-based conflicts.

Phenomena like social loafing and conformity/conflict have significant derivative effects on group dynamics. Derogatory terms like "slacker" and "overachiever" reveal the tensions that can be created in a group over social loafing. Effective group performance requires a foundation of common ground, that is, shared knowledge about local context, conventions, and co-reference to enable efficient and reliable interactions. Sustained group performance requires the development of trust and generalized reciprocity, sometimes called *social capital* (Coleman 1990).

Many of the challenges of collaboration are inherent tradeoffs; they can be addressed, and perhaps balanced, but not solved tout court. For example, designating a "coordinator" to receive and direct all group communications can improve group problem solving efficiency, but decreases satisfaction with the group activity (Leavitt 1951). Similarly, including a "skeptic" in brainstorming allows groups to produce more and better ideas, but also decreases members' satisfaction with the group activity (Connolly and Valacich 1990).

These collaborative challenges are as old as human organizations, but they are exacerbated by the very nature of knowledge work like software design and development. In knowledge work the interim work products, sometimes even the final work products, can be quite insubstantial. They are plans and strategies, architectures, algorithms and heuristics. The products of knowledge work are also typically arcane; indeed, software systems are possibly the best example there is of this.

8.3 Collaboration Supports Rationale

Collaboration is an important social resource for design rationale. The collaborative interactions of various software professionals ineluctably and naturally externalize rationales, though often incompletely. Collaborative interactions in software development also shape the software development process in ways that favor rationale.

8.3.1 Collaboration Externalizes Rationales

The creation of design rationale is often conceived of as a documentation activity within the software development lifecycle. And it is certainly true that design rationale can be a kind of documentation. Incorporating rationale into formal documentation activities is useful and efficient, since rationale provides causal foundation for other categories of documentation such as final specifications, reference and maintenance manuals, and user documentation like online help and tutorials.

But rationale is more broadly the reasoning that occurs throughout design and development, *whenever and however it is codified and used*. One of the most important consequences of collaborative work is that co-workers must articulate and externalize knowledge, assumptions and reasoning that otherwise might remain tacit. If you watch one programmer at work, you most likely would get little insight into programming. The work activity is mostly mental, and the occasional external inscriptions that are produced are quite arcane. But if you watch *two* programmers collaborating, you see quite a lot about programming. More specifically, you see quite a lot of rationale.

Software development is a complex, intellectual task in which there are never singularly correct solutions. More typically, there are many satisfactory solutions, each entailing a variety of partially understood tradeoffs and side-effects. Elsewhere in this book we have characterized these problems as wicked (Rittel and Weber 1973) or ill-structured (Reitman 1965) When people work on this kind of task collaboratively there is lots to talk about, indeed, lots to analyze, justify, and debate.

As Kraut (2003) put it, this kind of collaborative work follows a "trust-supported" heuristic in which group performance can be only as good as the second best member. Groups pool and weigh different perspectives; they identify and repair errors in candidate solutions and the rationale for candidate solutions. Producing a solution requires both the technical enterprise of identifying and developing a proposal, but also the social enterprise of convincing one's colleagues.

An old chestnut of software engineering is that no one wants either to produce or to use documentation. But in collaborative contexts, in which one must obtain the support of colleagues in order to make a technical decision, there is no shortage of design rationale. Indeed, the culture of software development work has evolved a variety of mechanisms to capture, preserve, and discuss these materials, such a commenting and literate programming (Knuth 1970), bug reports and frequently asked questions (FAQ) forums, and indeed the entire spectrum of Usenet communities.

Collaboration in software development unavoidably and voluminously generates rationale.

8.3.2 Software Development Communities of Practice

Software development is diverse and somewhat fragmented as a profession, but it is a profession that is all about the skills and practices of constructing software. Software professionals have developed a *culture* of software development – communication and work practices to coordinate work and to teach and coach one another (Curtis et al. 1988; DeMarco and Lister 1999; Lammer 1988). For example, software developers frequently talk about software components and their interactions in explicitly anthropomorphic terms; thus, a component is said to *know* things – such as how to put a file on the print queue – or to *expect* things from other components (e.g. Herbsleb 1999; Madsen 1994). In this sense, software development is a *community of practice* (Lave and Wenger 1991).

One could regard the cultural practices of software developers as curiosities. But in fact social practices emerge, evolve, and persist because they add something to human activity. Thus, it seems prudent to consider how the ways software professionals talk about and construct software – particularly those work practices that are *not* taught in formal education or encouraged by industry standards, corporate policies, or managerial directives – may reveal important characteristics about experts think about software, and how they coordinate software development work.

In this light, consider the issue of anthropomorphic and other metaphorical language. Formal education and normative practices in software engineering have traditionally placed high value on explicit and correct representations such as specification languages, programming languages, and a variety of diagrams. Notably these formal representations are pretty much strictly declarative; they describe the structures and interactions in a software design and implementation. Classic articles on computer science education by Dijsktra (1989), among others, have specifically argued against metaphorical language.

Why then would software developers employ anthropomorphic and other metaphorical language? Carroll and Mack (1985) argued that metaphorical representations clarify new domains by leveraging concepts that are already known, while at the same time highlighting mismatches in the mapping of old-to-new, and thereby flagging conceptual problems that need attention. Rosson and Alpert (1990) suggested that the anthropomorphic metaphors of object-oriented design facilitate upstream communication among developers by reducing the need for explicit point-by-point clarification and refinement entailed by more explicit representations. For example, saying that software component A knows about software component B is both succinct and rich. It conveys that the behavior of A depends in some way on the behavior of B, and that the specific nature of the contingency is either not yet known or not needed for present purposes.

Herbsleb (1999) elaborated this conjecture by noting that the strategy of anthropomorphic representation allows software developers to leverage "naïve psychology" (Clark 1987) – the near-universal understandings that humans share about animate entities. Naïve psychology allows people to reason what an animate entity must have known to have acted as it did, or what it is trying to do given its behavior and knowledge. In other words, it bundles declarative understanding of what is happening with direct perception of its rationale – that is, *how* an entity is able to do what it does and *why*. It is believed that naïve psychology capabilities were selected in evolution because individuals who could draw these inferences were better able to succeed in the early social world (Clark 1987).

Herbsleb (1999) analyzed a corpus of 1800 system behavior descriptions identified in a series of software engineering domain analyses. The domain analyses involved teams of 3-5 experts analyzing message-passing protocols in telephony or switch maintenance software. Herbsleb found that 70% of the behavior descriptions were metaphorical. Each domain analysis involved a series of meetings; for each series, Herbsleb analyzed one early meeting, one meeting from the middle of the series, and one of the final meetings. He found that through the course of the three domain analysis meetings, teams of software engineers came to rely *increasingly* on certain of these metaphorical descriptions – the ones derived from naïve psychology. That is, metaphors were not used as ephemeral ice-breakers, to replaced with more proper and explicit descriptions. Instead, they became established in the domain analysis as a sort of local technical language for the teams.

Communities of practice are social mechanisms for the codification and social transmission of practices and their rationales. Collaborative software development work requires sharing extraordinarily complex information fluently. Software development has evolved as a community of practice to leverage naïve psychology via anthropomorphic metaphors selectively hiding and emphasizing information, while bundling description and rationale. Another example of this in contemporary software practice is pattern languages (Gamma et al. 1995). Both simplify and speed communication in software collaborations by leveraging rationale.

8.4 Rationale Supports Collaboration

The relationship between rationale and collaboration is reciprocal. The role of rationale in software development is motivated and facilitated by collaborations among professionals. But rationale also supports collaboration. It provides a compelling management tool for keeping projects on track spanning time, distance, and organizational change. It facilitates awareness of one's team members, contributing to the development of common ground and trust, and it facilitates coordination, particularly in project contexts of high uncertainty.

8.4.1 Awareness

In order to collaborate effectively in a large and complex project one needs to know many things about one's collaborators (Carroll et al. 2006): Who are they? What do they want to do? What are they doing now? What tools are they using? To what other resources do they have access? Who do they work with? What are they thinking about? What do they know? What do they expect? What are they planning to do in the near future? What sorts of significant relevant experiences have they had in the past? What disciplinary biases and assumptions do they bring to this interaction? What do they value? What criteria will they use to evaluate joint outcomes? How is their view of the shared plan and the work accomplished evolving over time?

This may seem like a long list, but in fact it is quite incomplete. Consider the issue of coordinating nuances in vocabulary. A user interface designer and a software architect may both support prioritizing design elegance; they may even be able to talk at length about how and why this objective is important. But in practice, they may have entirely different notions of what elegance is. If goals are not adequately analyzed and codified, this kind of failure of common ground can quickly lead to conflict, putting the collaboration and the project outcome at risk.

Of course the mere fact that different professional perspectives differ with respect to technical concepts and skills, values and priorities, and so forth is *not* the problem. Indeed, such differences are required for a successful large-scale software project, or any other large-scale human endeavor. Professional diversity can be, should be, and often is a resource to a software development team. The challenge is to efficiently recognize and effectively manage these differences.

This is where rationale can help. To the extent that people share concepts, skills, values and priorities, they can more easily create and develop common ground and trust. This is the essence of a *community of practice* (Lave and Wenger 1991), as discussed earlier in this chapter. When team members *do not* share disciplinary concepts, skills, values and priorities – as in the example of the user interface designer and the software architect discussing *elegance*, they need to socially construct common ground by exchanging perspectives and attaining mutual understanding. (Analogous points could be made for social structures other than discipline and community of practice, such as culture and ethnicity). Members of a software development team can construct common ground by sharing their goals and visions for a project, their ideas about how to turn these into plans and actions, and what they most value and what they think they can contribute to the project.

A great variety of groupware tools are being developed, deployed, and investigated to provide awareness support in collaborative work – for example, tools for online discussion about, or direct annotation of project objects, various activity visualizations, personal profiles and social networks, and activity integrators (Carroll et al. 2006). All of these tools help to codify bits of rationale; many have the effect of making personal rationales more permanently accessible to other team members, or more closely integrated with project data objects. They help people share more of their reasoning and their reasons with one another, and that helps them collaborate more effectively.

8.4.2 Coordination

In their empirical study of collaborative coordination in large software organizations, Kraut and Streeter (1995) found that *informal discussion among team members* was the both the most valued and the most used coordination technique among the 18 coordination techniques they studied. Curiously, they also found that members of the software teams they studied valued informal interaction more than they actually engaged in it. More generally, Kraut and Streeter found that less formal coordination mechanisms – such as group meetings, discussions with one's manager, requirement reviews, design reviews, and customer testing – mechanisms that bring to light diverse viewpoints, were judged as valuable given the extent to which they are used, whereas more formal coordination mechanisms – like status reviews, code inspections, CASE tools, data dictionaries, milestone schedules, and source code – were judged as not valuable relative to the extent to which they are used.

In this investigation, the rated importance of informal and social coordination mechanisms in large software projects was strongest during periods of high uncertainty, such as in requirements and early design. In other words, the easy exchange of rationale, facilitated by less formal coordination mechanisms such as meetings and discussions, was critical to collaboration among software developers in high uncertainty, upstream stages.

Kraut and Streeter also noted a somewhat alarming tendency for projects to increasingly *de-emphasize* informal interaction through the course of development. Managers tended to prefer formal coordination mechanisms, and to shift towards these as possible.

Kraut and Streeter concluded that an important potential advantage in software management would be to devise better tools and techniques to enhance informal and interpersonal communication among team members throughout the development process. They note that many of the most prominent and celebrated techniques in software engineering, such as formal specification languages, are designed to minimize interpersonal communication. These may satisfy a managers' desire for formal coordination mechanisms, but they do not facilitate the easy exchange of rationale and were rated by developers as valueless relative to their use.

8.5 Summary and Conclusions

This chapter surveys the relationship between collaboration and design rationale. First, it observed that software development is almost always collaborative work, for the simple reason that most software projects are too big for solitary individuals to ever successfully tackle. This raises a set of specific challenges: Collaboration aggregates individual efforts, but it also creates new sources of work for people in teams, and new risks for the products of team work. We then consider how collaboration supports rationale in software development – by encouraging team members to explicitly articulate their goals and plans, and therefore to create the possibility of a discussion about reasons, and by supporting a culture of software development to conventionalize and leverage social mechanisms like anthropomorphic metaphors and software patterns. Finally, we consider how rationale supports collaboration in software development – by supporting awareness of how the project is meaningful to one's collaborators, and coordination among collaborators, especially with respect to making progress in uncertainty.

9 Change Analysis

Keeping track of how changes in decisions require changes in other decisions is crucial in design and development. By capturing decision tasks and decision alternatives in the rationale, and by recording the dependencies between these decisions, we can help to anticipate the effects of changes and identify the different kinds of inter-decision dependencies. In this chapter we explain the implications of change analysis for rationale usage and rationale support systems in software engineering.

9.1 Introduction

9.1.1 Issues with Change in Software Development

In any successful system, change is inevitable. This is expressed in Lehman's first law of Continuing Change: a program that is being used in the real world will need to change or will become less and less useful in that real world environment (Lehman 1996). Changing a software system, particularly one that is in operation, is a difficult proposition. There are several questions that are raised when a change is proposed:

- How will the change impact the software and related artifacts (requirements, design, code, testing)? What are the likely costs and risks involved?

- Is this change consistent with system requirements (functional and nonfunctional)? How will the consistency (or inconsistency) be managed?

These suggest approaching change analysis from two directions: impact assessment, where the change is analyzed to determine the extent, cost, and risk of the change, and consistency management, where the change is analyzed to determine if it is consistent with the requirements and goals for the software system. Consistency management also includes managing inconsistency in a system. While inconsistency is undesirable, there are cases where it may be best to leave inconsistency in place to avoid committing to its resolution too early (Nuseibeh et al. 2000).

Rationale can assist with impact assessment by linking requirements, decisions, and implementation. Rationale can assist with consistency management by providing the intent behind earlier decisions that should serve as a basis for the consistency checks. Rationale also supports change analysis by providing a means for documenting the intent of the changes themselves.

The amount of support that can be provided by the rationale depends on both its availability and structure. One example, which we will use to illustrate our points, is the RATSpeak notation defined for the SEURAT system (Burge and Brown 2006). The SEURAT rationale uses a semi-structured argumentation format that captures four types of arguments for and against decision alternatives: arguments referring to requirements, arguments referring to claims (non-functional requirements), arguments referring to assumptions, and arguments describing dependencies between alternatives. Figure 9.1 shows a portion of this argumentation structure.

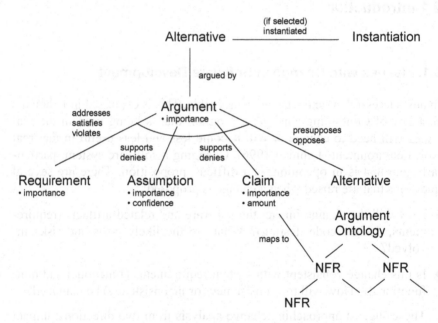

Fig. 9.1. Rationale Argumentation Structure

This is one example of a semi-structured argumentation format and is shown here to illustrate the types of arguments that are considered when making software decisions.

9.1.2 Objectives of this Chapter

This chapter describes the sources and types of changes made to software over the course of a system's lifetime and how these changes can be assisted by rationale. The chapter then focuses on two issues where rationale can support software change: change impact assessment and consistency management.

9.2 Types of Software Changes

There are many reasons why software requires modification. Changes come from an initial source, or cause. Reasons for needing to make a change can be broken into several, not mutually exclusive, categories:

- *Requests* – a change can be requested by the (or a) customer, system users, or management.
- *Defects* – a change can be a response to a defect in the system.
- *Operational Environment* – a change can be due to some change in the environment in which the software is operated. This would include changes in hardware configurations and changes in any laws/policies that would affect the software while in operation.
- *Development Resources* – a change could be due to a change in resources available for future development of the system. This would include changes in personnel available to work on the project and changes in COTS or other component suppliers.
- *Political Environment* – this could refer to the environment at the development company or the client company.

The proposed change, which often requires retracting a previously made decision for a decision task and selecting (or proposing) an alternative decision, needs to be analyzed to determine if the change affects the functional requirements (added functionality, changed functionality, or removed functionality), non-functional requirements (quality goals for the system have been changed), assumptions used in making development decisions, the structure of the system itself (adaptive or preventative maintenance changes required for future enhancements), or require fixes to defects in the software that do not derive from errors in requirements.

Part of the change analysis process should also involve an assessment to determine if the change should be performed, deferred, or rejected. The change may conflict with current system requirements in a way that can not be resolved. A requested change may be desirable for some system customers or users and undesirable for others. The potential cost of the change must be weighed against the expected benefit. The change may be evaluated as necessary but deferred to a later date. The arguments for and against making the change can be captured in the rationale for the request. This is especially crucial for changes that are rejected—it is not unlikely that the issue will be raised again in the future and having the reasons for the decision available will be crucial in determining if the initial rejection is still valid or should be re-considered.

9.2.1 Functional Requirement Change

Functional requirements may require change for many reasons. Requests for additional functionality are a common type of change and adding, or enhancing, system functionality results in new requirements. Requirements may also require modification. There may be cases where the original requirement needs to be relaxed or strengthened. The requirement may have initially been ambiguous or incorrect, resulting in software defects.

The rationale can assist in requirement change in several ways. For additions, the rationale for the additions should be captured to assist future developer/maintainers if the requirement needs modification in the future. Implementing the new functionality will require decisions on how that implementation should be done. The rationale for the design and implementation alternatives that are proposed should refer to the new requirements and to any existing requirements affected by the proposed changes.

For modified requirements, the rationale for the requirement will provide the intent behind the initial version of the requirement. This information should be taken into consideration in determining if and how the requirement is modified. The rationale can also provide traceability to decisions made during design and implementation that were made based on the original requirement. This can point out places where these decisions should change. The requirement may also be associated with arguments for and against alternatives considered and rejected when building the system. If the requirement has been changed, some of these alternatives may merit reconsideration.

Traceability from requirements to decision alternatives, as captured in the rationale, can be used to determine the effect of removing an alternative. Choosing a notation that supports this can then be used by a Rationale

Management System to indicate which alternatives are argued by the requirements and by re-computing alternative support if a requirement is removed or disabled and reporting if this action should result in reconsidering previous alternative selections. One example of a system that utilizes this information to predict change impact is SEURAT (Burge and Brown 2006).

Rationale should also be captured for the change. This provides a history of how the software has been modified over time and for what reasons. An evolution history can be used to determine where problems have frequently occurred in the software (Nierstrasz 2005) and to predict what future evolutions may be needed (Antón and Potts 2001). The ability to capture the historical information provides process-oriented rationale (Conklin and Burgess-Yakemovic 1991) where the focus is on using rationale to capture a history of the design process rather than a representation of the "design space."

9.2.2 Non-Functional Requirement Change

Non-functional requirements are qualities or characteristics that are desirable for the system being developed. They are also known as the "ilities" (Filman 1998) and include scalability, reusability, maintainability, etc. NFRs can be viewed as characteristics that constrain how, or how well, the system provides its functionality.

The importance of an NFR may change over the lifetime of a system. This could be in response to a change in how the software is used. For example, the number of expected users may increase, necessitating a stronger focus on the scalability of an application. Change may also be needed based on user feedback. If the users are unhappy with the performance of an application then any decisions made that affect system performance may need to be re-considered.

If the rationale has captured the impact of NFRs on design and implementation decisions, the rationale can be used to evaluate the impact of changing NFR priorities. In SEURAT, NFR priority, or importance, can be changed and used to recomputed support for alternatives captured in the rationale (Burge and Brown 2006). The NFR Framework (Chung et al. 2002) was used in the Goal Centric Traceability (Cleland-Huang et al. 2005) approach. Impact analysis was performed using a SoftGoal Interdependency Graph (SIG) by propagating changes made to goal contributions through the graph. The graph is linked to the functional model captured in UML.

Rationale should also be captured for changes in NFR priorities. This can be used to analyze how these priorities changed over time to help predict future changes to the current system and potentially to predict changes to other, similar, systems.

9.2.3 Changing Assumptions

Software decisions are often based on, or at least influenced by, assumptions. Unlike requirements, assumptions are not entities that must be true about the developed product but are entities that are *believed* to be true about the environment in which it must operate. An assumption may be made because developers are working with incomplete information or information that they are aware is likely to change over time. Some assumptions have a strong temporal component, as shown in rationale for a spacecraft design where designers made decisions assuming that certain technologies would be available in the future when the spacecraft was actually built (Oberto 2002).

The gradual change in validity of assumptions over time is a key driver of software evolution (Lehman 2005). This indicates a need to be able to easily assess their impact on the software product in order to respond to, or even anticipate these changes. This can be supported by capturing these assumptions, and their role in software decision-making, in the rationale. In cases where an assumption is known to have an "expiration date", such as the spacecraft example mentioned earlier, it should be possible to use the rationale to alert the developer/maintainers when that date approaches so they can re-evaluate these assumptions to determine if they still hold (Burge et al. 2006). When an assumption is known to no longer be valid, the rationale can be used to determine its impact on the system by viewing its relationship with selected alternatives. The removal of the assumption can also be used to re-calculate alternative support and alert the developer/ maintainer if the change in assumptions should require reconsideration of alternative selections (Burge and Brown 2006). A more prescriptive approach to assumptions is taken by the REMAP system (Ramesh and Dhar 1994). In REMAP, the rationale is used to help the user select from design object alternatives in a design library. If an assumption that guided an earlier choice is retracted, the affected design object is automatically retracted and a new one is selected by the system.

If new assumptions are involved in system changes these assumptions should be captured in the rationale so that this information will be available to use if the assumptions no longer hold in the future.

9.2.4 Structural Changes

Sometimes software changes are required in order to make the software more maintainable. These changes are referred to as perfective (IEEE 1998) or preventative (Lientz and Swanson 1988) maintenance. Such changes are often necessary as evolving systems become increasingly complex (Lehman 1996) over time. These changes could be refactorings to remove "bad smells" in code (Fowler and Beck 1999) or major system reengineering efforts (Sommerville 2007) where much of the application is re-written. Rationale can support both these efforts.

When code is modified, the presence of the rationale associated with it can be used to provide the author's intent. This additional insight can help prevent defects from being introduced due to misunderstandings of the original implementation. One of the authors of this book experienced this first hand when a coworker made a software change that removed what they thought was a defect in the code and introduced (or more likely, re-introduced) a timing error that damaged the hardware the code was controlling.

Rationale is also useful in documenting the structural changes and the reasons behind them. Design Patterns (Gamma et al. 1995) can be of great assistance in writing code that is more easily extensible but if patterns are not documented, may be "broken" in subsequent modifications to the code, negating their value. Relating the actual changes to the reasons for making them can also assist with traceability if structural changes lead to defects.

9.2.5 Defect Correction

Some software changes are in response to defects in the software discovered during testing or operation. The defect may have arisen from misinterpreting a requirement, making a poor design or implementation choice, or be a simple coding error.

In the first two cases, the rationale can provide a link to where the defect may have been introduced. This is of considerable assistance since finding the relevant code can take a significant portion of the repair time (Ko et al. 2005). If the problem can be traced to a misinterpretation of a requirement, the rationale can be used to determine what decisions were based on the presence of the requirement and point the way towards the implementation. Similarly, if the problem was introduced by a bad decision, the rationale will indicate where the decision was implemented and also what some alternatives might have been. For coding errors, rationale associated with the code is useful in explaining the relevant code and may help to

prevent modifications that conflict with the developers' intent and introduce additional defects.

9.3 Change Impact Assessment

When a software change is proposed, there are two decisions that require
impact assessment. The first is if the change should take place at all. It is
possible that the proposed modifications may conflict with other system
requirements and may not be beneficial to the majority of system users. It
is also possible that the proposed modifications maybe desirable but not
enough to outweigh their costs. The second decision is the decision of *how*
to make the modification. There may be more than one alternative that
should be considered.

Both these decisions require some level of impact assessment to determine what affect the proposed change will have on the current system.
This assessment can be made more quickly and accurately if the maintainers are given the rationale behind the development decisions. This was
shown in a study that compared impact assessment using only source code,
standard documentation, and model dependency descriptors (Abbattista
et al. 1994). This study showed that using the model dependency descriptors
(Cimitile et al. 1992), which contained rationale, increased impact assessment accuracy.

There are many approaches to determining impact at the code level.
These include analyzing the source code call graph (Bohner and Arnold
1996), static and dynamic program slicing (Weiser 1981) (Agrawal and
Horgan 1990) (Gallagher and Lyle 1991), path-based impact analysis (Law
and Rothermel 2003), program change histories (Canfora and Cerulo 2006)
(Zimmerman et al. 2004), and approaches that use multiple techniques
such as the Technical Risk Estimation (TRE) tool which uses dependency
structure and change history (Walker et al. 2006) to predict how change
will propagate. These approaches evaluate the impact starting with a set of
predicted code changes or by detecting similarity to prior changes.

Another approach to impact analysis is to provide traceability between
related artifacts through meta-data describing their relationships. An example of this is the software repository developed by Sneed (2001). This
repository captures concept models, code models, and test models. The relationships captured in these models are used to determine the impact of a
proposed change.

Rationale is another form of meta-data about the software and can
also support impact analysis. Most impact analysis techniques require

identification of a set of "trigger" objects (Queille et al. 1994) that indicate where the change will start. This can be difficult if the change originates at a high level such as a change in NFR priority or the invalidation of a key assumption. The rationale mapping the NFRs, requirements, and assumptions can assist with detecting which objects are involved. The rationale can also be used to detect similar changes that occurred in the past by comparing rationale for the currently proposed change and the past changes.

9.4 Consistency Management

One of the challenges in developing and evolving large software systems is to maintain consistency when possible and to manage inconsistency when consistency is not possible. There are numerous approaches toward software consistency that look at consistency in requirements (Klein 1997), code (Tarr and Clarke 1998), views/perspectives (Finkelstein et al. 1994; Grundy et al. 1998) and between software artifacts (Riess 2002; Nentwich et al. 2003). These approaches utilize constraints to check for consistency.

In some cases it is not possible, or even desirable, to eliminate inconsistency. Tolerating inconsistency may be necessary if inconsistencies are too expensive to repair, if the information required to resolve the inconsistency is not known at the current stage of the development, or if it is too early in the process to make the design decisions required for resolution. The inconsistency can be ignored, and revisited at a later date, deferred until a later time, circumvented by changing the rule that indicated the consistency was present (if the inconsistency is an exception or if the rule is incorrect), or partially resolved (Nuseibeh et al. 2000).

Software engineering rationale provides another tool in consistency and inconsistency management. While other tools look for inconsistency between developed artifacts, rationale can also be used to detect inconsistencies in the decision-making process and the developers' reasoning. For example, semantic inference is performed over the rationale captured in InfoRat (Burge and Brown 2000) and SEURAT (Burge and Brown 2006) to look for selected alternatives where their arguments were contradictory. In SEURAT, NFR priorities can be assigned at a global level and propagated through the rationale to evaluate alternatives and report if selections were inconsistent with overall system goals. If the global priority is not applicable to a specific decision it can be overridden and these overrides are saved so they can be reported on if necessary.

Results of rationale-based consistency checks do not have to be resolved immediately. In SEURAT, they are reported as warnings and can be overridden if necessary. As with priority overrides, this information is stored so the override can be removed later if necessary. Rationale can also support the ability to record questions that come up during the decision-making process that need to be resolved before the decision can be made. These questions support inconsistency management by providing the developers with the means to explicitly indicate where more information is required before an inconsistency can be resolved. The ability to capture questions and information on potential methods for their resolution is supported by the Decision Representation Language (DRL) (Lee 1991) for use in SIBYL and was also implemented in SEURAT (Burge and Brown 2006).

9.5 Summary and Conclusions

Software systems undergo many different types of changes during their lifetimes for a variety of reasons. Change comes with risk: risk that the change is incompatible with decisions made earlier, risk that changes are implemented incorrectly or incompletely, risk that change introduces inconsistency into the system. These risks can be mitigated by the presence and use of rationale. The rationale describes decisions made earlier and the intent behind the developers' choices. This information is invaluable when these decisions change and can help to prevent problems such as repeating past mistakes, introducing conflicts with earlier choices, and using reasoning that is inconsistent with earlier efforts.

In this chapter we described how rationale can be used to support the different types of changes made to software systems and how it supports two key aspects of change analysis: change impact assessment and consistency management. The success of making a change to a software system is directly affected by the depth of knowledge the modification is based on. This knowledge is greatly enhanced by rationale that indicates not only what the system does but why.

Part 3
Rationale and Software Engineering

The importance of rationale in software engineering is underscored by rationale being featured as a key activity in recent talks on the Future of Software Engineering (Taylor and van der Hoek 2007;Whitehead 2007) and by rationale being featured as part of one of the process areas in the Software Engineering Institute's Capability Maturity Model Integration: Decision Analysis and Resolution (CMMI Team 2006).

Decisions are made all throughout the software development process ranging from deciding how customer requests can be translated into software requirements to deciding when and how to adapt software in operation and on to when a system is ready for retirement (Chapters 10 through 14). The rationale behind those decisions documents the developers' intent and keeps this information from being lost forever due to attrition, reassignment, or by simply being forgotten.

An important aspect to software development that cross-cuts development phases is re-use (Chapter 15). As software increases in complexity and cost, it becomes critical to avoid "reinventing the wheel" and to utilize existing software applications to save time, by buying instead of building, to save money, since the price to purchase an off-the-shelf application is often less than building it yourself, and to increase reliability, by working with applications that have already received extensive evaluation. Still, while re-use can potentially meet these valuable goals, it is not without its dangers. Deciding when and how re-use should be utilized, and what the best candidates for reuse are, must be carefully deliberated.

The ability of a software system to fulfill the needs of its stakeholders is directly dependent on degree to which those needs were taken into account by the developer. By capturing the decisions made during development and relating the alternatives chosen to the stakeholder needs, it is possible to use this Software Engineering Rationale to assess the ability of the software to meet those needs. This is the essence of Rationale-Based Software Engineering.

10 Rationale and the Software Lifecycle

Software development can be modeled using a number of different lifecycle, or process, models. These include the waterfall model, the spiral model, the Unified Process, V-Model, and others. In this chapter, we will describe these models and how rationale capture and use supports the development process followed in each of them.

10.1 Introduction

10.1.1 Software Engineering Process

The software engineering process and the software lifecycle are closely related concepts. The software lifecycle refers to the stages of software development that take place over the lifetime of the software. The IEEE/EIA defines the primary lifecycle processes to be acquisition, supply, development, operation, and maintenance (IEEE/EIA 1996). There are also supporting processes and organizational lifecycle processes (IEEE/EIA 1996). Supporting processes include documentation, configuration management, quality assurance, verification, validation, joint review, audit, and problem resolution. Organizational lifecycle processes include management, infrastructure, improvement, and training. While the ISO/IEC standards described earlier take a high view, the most typically mentioned lifecycle stages encompass the development and maintenance lifecycle processes and include requirements analysis and specification, design, implementation, integration, verification and validation (testing), installation/deployment, maintenance, and retirement. Software lifecycles are modeled by a variety of software process models that define how the development stages progress. The lifecycle model defines the "skeleton and philosophy" of the process (Fuggetta 2000).

The software process is what controls and monitors the development described by the lifecycle model. The software process is defined by Fuggetta (2000) to be "the coherent set of policies, organizational structures, technologies, procedures, and artifacts that are needed to conceive, develop, deploy, and maintain a software product."

Rationale can play a role in software process by capturing the reasons behind both process and product decisions. The product rationale captures the reasons for decisions that directly impact the delivered product while the process rationale describes the reasons behind the process selected to guide the product development. Process decisions are important because the process chosen needs to fit the size of the project, the experience level of the development team, and the development tools available.

10.1.2 Objectives of this Chapter

In this chapter, we describe the stages of the software development lifecycle and how rationale applies to each of them. We also describe a number of software lifecycle models. We conclude with a section on software process improvement.

10.2 Development Activities and Rationale

The software life-cycle consists of a number of stages of software development. In this section, we briefly describe a typical set of development stages and how rationale can be captured and used in each of them.

10.2.1 Project Planning and Management

While project planning and management is listed first among the stages, planning and management are ongoing activities throughout the development process. Project planning involves many decisions: delivery date, staffing needs, budget, milestones, deliverables, etc. These decisions involve many tradeoffs. For example, one tradeoff might be assessing the importance of short time-to-market vs. the amount of functionality provided or the quality level of that functionality (how much time to spend on validation and verification). These decisions and the reasons for the choices made should all be captured in the rationale. The process of recording deliberation during planning as rationale assists with collaboration and negotiation.

Management decisions can also be captured in the rationale for the project. The use of rationale in software project management is described later in this book. Rationale can support collaboration, risk management, success criteria reconciliation, process improvement, and knowledge management.

10.2.2 Requirements

Requirements engineering is arguably the most crucial stage in the software lifecycle. Failing to adequately capture and refine requirements is considered to be a leading cause of project failure (Alford and Lawson 1979; Hofmann and Lehner 2001). Rationale can support requirements elicitation by capturing reasons behind requirements and allowing comparison with stakeholder needs; requirements negotiation by capturing the deliberation process; inconsistency management by allowing comparison of priorities across requirements; and requirements prioritization (a key element in Value Based Software Engineering (Boehm 2006a)) by associating priorities to the criteria behind each requirement, both functional and non-functional.

Rationale can also play a large role in requirements traceability by providing the means to associate the decisions made later in the development process with the requirements that drive them. This applies to both the functional requirements as well as non-functional ones. Both types of requirement can appear in arguments for and against alternatives that are captured in the rationale.

10.2.3 Design

Much of the research involving rationale has been in the area of Design Rationale – the reasons behind design decisions. In software, there are several levels of design that take place depending on the size of the system being built. High level design is often referred to as architectural design. This stage involves designing or selecting, the software architecture. The choice of architecture is often driven by the "quality requirements" (non-functional requirements) of the system. For example, Attribute-Based Architectural Styles (ABAS) (Klein and Kazman 1999) associate software architectural styles with quality attributes such as performance, availability, and modifiability.

The design process progresses from the high-level decisions made when performing architectural design into the lower-level decisions in detailed design as classes, or modules, are designed. The rationale can be used to capture the decisions made at this point in the process and eventually linked to the code that will implement the alternatives selected.

10.2.4 Implementation

Implementation involves translating the design into the executable source code. There are still decisions made during this part of the process and the rationale for these decisions should be captured. The rationale can be evaluated to ensure that the reasons chosen are consistent with those given at earlier stages of development. The rationale can also be used during software maintenance to describe why the software was implemented the way it was and to help prevent new decisions from counteracting those intentions.

10.2.5 Verification and Validation

In order to insure that the developed system provides the functionality needed by the customer and that it meets its specification, it needs to be tested. The evaluation process is typically described as verification and validation, or V&V. While we often describe this stage as occurring after implementation, in reality V&V activities should take place all the way through the development process. Test planning should be started when the project planning is performed, requirements should be examined to ensure that they are testable, unit testing should be performed during implementation, system testing is performed prior to deployment, and regression testing (as well as any new tests) must be performed when changes are made during maintenance.

Boehm gave an often cited definition of the difference between validation and verification – validation asks "are we building the right product?" and verification asks "are we building the product right?" (Boehm 1979; Sommerville 2007). Verification involves ensuring that the software conforms to its specification while validation involves checking that the software does what the customer needs it to do.

Rationale can support software testing by providing insight into how quality factored into software decisions. This information can be used to determine where testing efforts should be concentrated. Collecting rationale for the testing effort itself would be useful in assisting with making testing decisions and in using the reasons behind testing choices and the results of these decisions to point out testing strengths and weaknesses that can be applied to future projects.

10.2.6 Maintenance

A successful software system is likely to require some form of maintenance over its lifetime. These changes can be challenging, especially if the original developers are not available. This is an area where rationale is especially valuable. Knowing the intent behind the decisions made when developing the software can help to prevent problems or inconsistencies being introduced during maintenance. If the rationale captures the assumptions made when initially building the system it can be used during maintenance to suggest where changes need to be made if those assumptions change. This assistance is provided in the Software Engineering Using RATionale system (SEURAT) (Burge and Brown 2006).

10.2.7 Retirement

If, or when, to retire a software system is potentially the last decision that needs to be made during the system's lifetime. The decision on whether to repair (maintain) or replace a system needs to be well thought out. This deliberation can be supported by and captured with rationale. The rationale for the decision would also be valuable if the retired system ends up being re-instated or re-used later.

10.3 Software Lifecycle Models

There are a number of different categorizations for software life-cycle models/software process models. Here we have chosen to break them into three categories: sequential models where development typically proceeds linearly through the phases, iterative models where iteration is built into the models, and a third category for models that do not fit into either of the two categories or that span categories.

10.3.1 Sequential Models

10.3.1.1 Waterfall Model

The waterfall model was originally defined by Royce (1970). In this model, development proceeds through the stages in a sequential fashion as shown in Figure 10.1. Each stage (shown as a box in the figure) needs to complete before the next stage can begin. The example shown here includes feedback loops indicating that it is possible to go back to make modifications to work done earlier if necessary. The stages vary slightly between different depictions of the model but typically include requirements, design, implementation, testing, and may also include maintenance, deployment, and retirement.

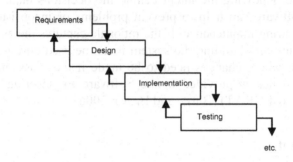

Fig. 10.1. Waterfall Model

The waterfall model has fallen somewhat out of favor. The separate stages are seen as being inflexible and less responsive to changing requirements. The model does, however, have the advantage that it is easy to assess where in the process a software project is, something not always clear with more iterative methods. This model resembles models used in other kinds of engineering projects and is often used when the software is part of a larger systems engineering project (Sommerville 2007).

Each of the stages captured in the waterfall model will include many decisions that will have a large impact on the later stages. Capturing the rationale for these decisions will help to ensure that decisions made in later stages will be consistent with earlier ones.

10.3.1.2 V-Model

The V-model is similar to the waterfall model but also includes the verification activities and how they relate to development stages. A key difference between the V-model and the waterfall model is that the level of abstraction is explicit (Bruegge and Dutoit 2004). Figure 10.2 shows a simplified V-model, adapted from (Bruegge and Dutoit 2004) (Jensen and Tonies 1979). As with the waterfall model, capturing rationale can help with the traceability of decision criteria throughout the process.

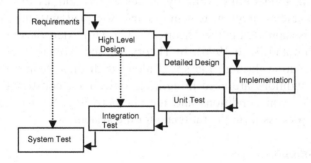

Fig. 10.2. V-Model

10.3.2 Iterative Models

Iterative models differ from sequential ones in that they depend on the software being built in a series of iterations. In this section we briefly describe some of the more common models.

10.3.2.1 Incremental Delivery

Incremental delivery consists of portioning the system into a series of releases. The initial requirement development and architectural design is done for the system as a whole but the functionality is delivered incrementally. This method has several advantages including making the software available to the users earlier, gaining experience with early increments to help refine requirements for later ones, reducing the risk of project failure, and ensuring that the most important functionality (typically developed in the earlier increments) receives the most testing (Sommerville 2007).

10.3.2.2 Spiral Model

The Spiral Model, developed by Boehm (1986), depicts the software development process as a series of increasingly more developed prototypes. The spiral moves through four quadrants. The first quadrant looks at objectives, alternatives, and constraints on the next development cycle. The second quadrant evaluates the alternatives proposed in the first quadrant and identifies and resolves risks. The third quadrant develops and verifies that level of the product (the prototype), and the fourth plans out the next phase or phases. This model both explicitly addresses risk and, by the alternative identification and evaluation steps in the first two quadrants, the rationale.

Rationale is supported in the Theory W (win-win) extensions to the spiral model (Boehm and Bose 1994). In Theory W, stakeholders are identified for each revolution through the spiral along with their "win conditions." These win conditions are used in defining objectives, constraints, and alternatives. The win conditions and the alternatives generated during the spiral model process form the rationale for the system.

10.3.2.3 Unified Process

Rational Unified Process (RUP) (Kruchten 1999) and its more general form, the Unified Software Development Process (Jacobsen et al. 1999), consists of four phases, with multiple iterations taking place during each phase. The four phases are inception, where the initial business case is defined; elaboration, where requirements and risks are defined; construction, where the system is designed, programmed, and tested; and transition where the system is moved into its operational environment (Sommerville 2007). Within each of these phases, there are nine core workflows: business modeling, requirements, analysis and design, implementation, test, deployment, project management, configuration and change management, and environment. The amount of effort spent in each of these workflows depends on the development phase. For example, more time is spent on business modeling and requirements in the inception and elaboration phases and less in the construction and transition phases. Similarly, the amount of implementation slowly increases in the first two phases, which may involve simple prototypes, while reaching its highest level in the construction phase when the actual system is built. The Rational Unified Process was developed by Rational Software and is supported by its products.

The Unified Process is a generic and comprehensive process that attempts to cover all aspects of software development. Because of its comprehensive nature, it can be seen as being too unwieldy for smaller development projects. The process can, however, be adapted to work with smaller projects (Hirsch 2002; Pollice et al. 2003). Process rationale can be captured to document how the process was tailored, and why. This information can then be used to transfer the lessons learned to future software projects using the same or similar processes.

10.3.2.4 Extreme Programming

Extreme Programming (XP) can be viewed as a variant on incremental delivery (Sommerville 2007). The extreme in extreme programming does

not indicate a "daredevil programming" but instead refers to taking existing best practices to the extreme (Beck 1999). The development process is
a collaborative one between the customer and the developer where functionality is described as a series of stories (similar to use cases) and where
each release chooses the set of stores that are viewed as the most important. Releases are developed using test-first development and pair-program-
ming.

   The goal of XP is to center the development process on coding and to
try to develop releases that are as simple as possible and to plan on refactoring later if necessary. The danger of this the difficulty of knowing where
there were short-cuts made that would need to be re-examined in later releases. Documenting the rationale for the decisions made in earlier iterations can be used to detect where alternatives were chosen in the interest of
expediency that may require change as requirements are added or refined.
The value of this is demonstrated by the Software Engineering Using RA-
Tionale (SEURAT) system (Burge and Brown 2006) where non-functional
requirement priorities can be modified and used to detect where earlier
choices should be re-considered. A rationale-based support system such as
SEURAT can be used during XP to detect candidates for refactoring.

10.3.3 Other Models

10.3.3.1 Rapid Application Development

The goal of Rapid Application Development, or RAD, is to build software
products more quickly, and with higher quality, than can be done using
more traditional software life-cycle approaches (Martin 1991). This is accomplished by taking advantage of Computer Aided Software Engineering
(CASE) tools and fourth-generation language tools. RAD is an approach
that can be used to build data-intensive business applications (Sommerville
2007) by exploiting commonalities between these systems: forms needed
for data input and display, database access, commonly used office applications such as word processors and spreadsheets, and report generation.
Many RAD projects are a form of COTS-based development projects because they link together existing Commercial Off-the-Shelf (COTS) applications to provide the required functionality (Sommerville 2007). RAD is
often confused with rapid prototyping but the key difference is that rapid application development is intended to build the final system while a

prototype is typically built to gain a better understanding of system requirements or available technology.

The success of a RAD development effort hinges on the selection of tools, products and COTS applications used in its construction. There may need to be compromises made to adjust system requirements so that they can be supported by these tools and components. Capturing rationale for the choices made and alternatives considered assists the selection process by making the reasons for selection and any tradeoffs made explicit. The rationale, and the alternatives captured in it, is also useful if subsequent versions of the system need to reconsider these decisions. RAD systems run the risk of dependence on third-party software where the vendor may go out of business, stop supporting the product, or raise licensing fees. These vendor changes may necessitate a change in the system to avoid problems.

10.3.3.2 Component Based Software Engineering

The Component Based Software Engineering (CBSE) development process builds software products out of re-usable components. The goal is to make software engineering more like other engineering disciplines where parts are ordered from a catalog and configured using well defined interfaces in order to create a new product. CBSE relies on the availability of components and on being able to adapt requirements, when necessary, to work with these components. CBSE is not strictly a process or a life-cycle. The components can be developed and used within any of the life-cycle models shown here.

Rationale can be used during CBSE by both component providers and consumers. For component providers, the component rationale can describe both functional and non-functional capabilities of the component. For component consumers, the rationale can be used to find a component that best matches the functional and non-functional requirements of the system under development.

10.3.3.3 Open Source Software Development

Open source software development involves multiple software developers working together over the internet to build software systems where the code is available freely to all. This has resulted in a number of successful software projects including the Linux operating system (www.linux.org), the Apache web server (www.apache.org), and Mozilla project products (www.mozilla.org) such as the Firefox browser and the Bugzilla bug-tracking system. There have also been open source projects with corporate support, such as IBM's Eclipse development framework (www.eclipse.org). The unifying attribute of these systems that has made them successful is that they are all systems that the developers want to be able to use themselves. Successful projects result from developers solving problems that they are excited about (Raymond 2001).

Since open source development is a highly collaborative process where developers can come and go from the project at will, the capture and use of rationale could play a significant role in the success of these efforts. Successful open source projects such as Apache and Mozilla make heavy use of version control systems, such as CVS, and bug tracking (Mockus et al. 2002). These systems capture the reasons behind software changes that could be included in their rationale. Capturing the intent behind the software modifications can be used to help guide the developers as the system evolves.

10.3.3.4 Model Driven Development

Models have been used to assist with software development for many years. The simplest definition of model driven development (MDD) is to built a model of a system that is then transformed into the system itself (Mellor et al. 2003). A more specific view is to develop domain models for application areas and use those to develop system architectures (Boehm 2006b). Models used in MDD can be developed using UML (France et al. 2006) or domain-specific modeling languages (DSMLs) that define relationships between domain concepts along with semantics and constraints (Schmidt 2006).

The usefulness of these models would be increased if they are developed with rationale attached. This would assist in selecting the appropriate model for the problem that the system is solving and could also help to determine when tailoring the model would be appropriate or not.

10.3.3.5 Service Oriented Development

In service oriented development applications are built using stand-alone services that can be executed on distributed computers (Sommerville 2007). Services are accessed via a service registry which is used to find applicable services. When a service is found by an application, the application is then bound to that service. A key aspect of service oriented development is the ability to perform "ultra-late-binding" where the service is located and bound dynamically (Turner et al. 2003). Web-services are an example of the service oriented development paradigm.

The uses of rationale in service oriented development are similar to those in CBSE: the rationale can be used as part of the selection criteria used when discovering service providers. For example, the Web services stack framework proposed in (Turner et al. 2003) includes a non-functional description level that provides a non-functional description of a service. These protocols would then provide the rationale for selecting the service.

10.4. Software Process Improvement

As described earlier, the quality of software products is related to the quality of the software process. In this section, we describe two process improvement initiatives: the CMM and CMMI process improvement framework and the Personal Software Process.

10.4.1 CMM

The Software Engineering Institute (SEI) developed the Capability Maturity Model (CMM) (Paulk et al. 1993) to define software maturity levels. These levels are initial, repeatable, defined, managed, and optimizing. At the initial level, the process is undefined and unpredictable. At the repeatable level there are policies and procedures in place for the software process. Companies working at the defined level have documented and standardized procedures that work across the organization. At the managed level metrics are collected to assess the quality of the software process and at the optimizing level this information is fed back into the process to improve it.

The Capability Maturity Model has been replaced with the Capability Maturity Model Integration (CMMI) (CMMI Team 2006). The CMMI integrates the software CMM with the Systems Engineering Capability Model (SECM) (EIA 1998) and the Integrated Product Development Capability Maturity Model (IPD-CMM) (SEI 1997). The CMMI has two representations – a staged model that assesses the organizations process at one of five discrete levels (similar to the CMM) and a continuous model where different process areas within an organization can be ranked at different capability levels. The capability levels are incomplete, performed, managed, defined, quantitatively managed, and optimizing. There are twenty four process areas defined within the CMMI. Examples are project planning, requirements management, and configuration management.

Rationale capture and use is related to the CMMI Decision Analysis and Resolution process area. This process consists of defining a "formal evaluation process" for evaluating decision alternatives. This process includes identifying the alternatives, determining the evaluation criteria, selecting and using the evaluation method, and selecting the alternatives based on the criteria (CMMI Team 2006). The evaluation process used on a project should determine which categories of decision will require formal evaluation (such as high risk decisions) and how the evaluation will be performed and documented.

10.4.2 Personal Software Process

The Personal Software Process (PSP) (Humphrey 1995) arose from applying the CMM to small software projects. The CMM focuses on improving the process of software development organizations and the PSP extends that focus to improving the process of individual software engineers. The PSP follows the principles that each developer needs to base their process on data that they collect on their own performance, the developers need to follow a defined and measured process, developers need to be responsible for the quality of their work, and that defects should be avoided if possible, fixed as soon as they are detected, and that the right way to do the job will be the fastest and cheapest (Humphrey 2000).

The PSP follows a process improvement cycle where the individual developers capture metrics on their job performance: time spent and defects introduced and removed. These metrics are then used to improve their performance. The PSP provides detailed forms and scripts to use during the development process.

The Team Software Process (TSP) (McAndrews 2000) extends the PSP to developing software in teams. The TPS addresses five causes of project failure: lack of training in planning, development, and quality practices; focus on schedule rather than quality; lack of a formal team-building process; unrealistic project plans damaging rationale. The TSP defines how Level 5 of the CMM can be put into practice.

Neither the PSP nor TSP calls for the recording of rationale as part of the process. The success of these approaches, however, indicates that emphasizing quality over schedule concerns leads to more successful projects. The addition of rationale to the collected data would add to this success by providing additional insight into the development process that can be used to then tune these processes during future development. It is clear from the results of PSP/TSP projects that spending time up front to collect data ends up improving the process and not having the detrimental effect on schedule that is so often feared.

10.5. Summary and Conclusions

The incentive behind the defining, modeling, and monitoring of the software life-cycle is to increase quality and decrease costs. Software process models have evolved from sequential models towards more iterative ones in order to be more responsive to changes in software requirements. The importance of a defined and monitored software process has been highlighted by process improvement efforts such as the CMMI and the PSP.

The capture and use of rationale should be an integral part of any development process. The usual software artifacts produced during development only describe what was done and not why. Knowing the information behind the decisions can provide much needed insight when these decisions are the basis of future ones. The reasons for making decisions that are captured in the rationale are often non-functional requirements that affect the overall software quality. The rationale can provide a way to evaluate that quality and support quality improvement.

Much of the opposition to the capture and use of rationale has been the view that it is difficult and time-consuming to collect. This argument can be used towards most forms of documentation but it is rare to find anyone who does not believe that documenting software will not save money in the long run. As software processes become more rigorous, the cost of collecting rationale will continue to become less of an issue compared to the savings provided by the defect reduction and requirement conformance provided by the improved processes.

11 Rationale and Requirements Engineering

Many of the decisions that have the greatest impact on the software development process are made during requirements analysis. Software Engineering Rationale (SER) can support this process by providing the ability to capture the decisions and the reasons behind them starting at these earliest phases. SER also supports requirements traceability throughout the process by directly mapping the development options chosen to the requirements that provide their rationale and by providing rationale for the requirements, thereby mapping requirements back to their source. In this chapter, we describe how rationale can support requirements engineering.

11.1 Introduction

11.1.1 Requirements Engineering

The key to every successful software project is its ability to meet the needs of its intended customer. This means that the software developers must determine what the requirements are for the software system. The process of identifying requirements, analyzing them to obtain additional requirements, documenting them in a specification, and validating that specification to ensure that it meets user needs is known as requirements engineering (Saiedian and Dale 2000). In provisioned systems (systems developed under contract), the requirement specification serves as the basis for the development contract; in product development, requirements are written based on market analysis and are expected to change if necessary (Kuusela and Savolainen 2000).

Inadequate or deficient software requirements are considered the leading cause of project failure (Alford and Lawson 1979; Hofmann and Lehner 2001). Lindquist (2005) states that analysts report the percentage of project failures resulting from poor requirements management as greater than 70%. The management problem is especially difficult on systems where the requirements are not stable. It is well known that the later in the development process a requirement changes the higher the cost to make the change will be. Agile development methodologies, such as Extreme Programming (Beck 1999), have been created to "flatten the curve" and be more responsive to changing requirements.

Requirements are typically broken into two categories: functional requirements that describe what the system should do (functions performed or features implemented) and non-functional requirements (NFRs) that describe qualities that the developed system should have. NFRs are often referred to as "ilities" (Filman 1998) since NFRs include qualities such as usability, scalablity, reusability, testability, maintainability, etc. Non-functional requirements are difficult to test and verify because they tend to cross-cut functionality of the system and also because they are often difficult to quantify. While they do not describe the functionality desired by the stakeholders they do have a direct impact on how satisfied the stakeholders are likely to be with the final product. Some NFRs involve the development process. Examples of these would be affordability, maintainability, and flexibility.

11.1.2 Objectives of this Chapter

This chapter discusses some of the key areas of requirements engineering and how they can be supported by the capture and use of rationale. In particular, the chapter focuses on obtaining requirements, requirements traceablity, approaches using non-functional requirements, goal-based requirements engineering and how rationale can support requirements change.

11.2 Obtaining Requirements

11.2.1 Requirements Elicitation

The first challenge faced in RE is the difficult task of eliciting requirements from the system stakeholders. Stakeholders are typically referred to as being anyone who is involved in the project or "whose interest the project affects" (Hoffman and Lehner 2001). This is a very broad category and can include the users, developers, marketers, procurers, QA, and possibly any others who might be affected by the use of the system. Sharp, et al. (1999) identify four groups of "baseline" stakeholders: users (those who interact with or control the software and those who use products of the system), developers, legislators (anyone providing guidelines for operation), and decision-makers (managers and finance people in both the developer and user organizations).

After stakeholders are identified, the next challenge is obtaining the requirements. There are many challenges encountered in this process which include stakeholders having difficulty expressing what they want or making technically unrealistic demands; stakeholders describing requirements in the language of their domain which may not be familiar to the analyst; conflicts in stakeholder requirements; political factors affecting requirements; and the possibility of the business environment changing (Sommerville 2007). The requirements specification can be viewed as a "wish list" for the different groups of stakeholders where the requirements rising from different stakeholder views may be inconsistent or contradictory (Kuusela and Savolainen 2000).

Requirements can be obtained using many methods. These include structured or unstructured interviews, observing the system in use (if the new system replaces an existing one), rapid prototyping to get user feedback, and collaborative approaches such as Joint Application Development (Bruegge and Dutoit 2000). The stakeholders may have a difficult time articulating their requirements. The more expert a user is at performing a task, the higher the chance that they will be performing at least parts of it "automatically," making it more difficult for them to describe those steps to another person. This necessitates a combination of direct and indirect elicitation techniques where direct techniques are used to obtain information that can easily be expressed verbally and indirect techniques are used to obtain information that can not be easily expressed verbally (Hudlicka 1997).

One thing that is typically not done during requirements elicitation is capturing the rationale behind the requirements. There may be many system features identified for potential incorporation into a software system. The rationale can capture the tradeoffs between these features along with the consequences, both desirable and undesirable of incorporating or not incorporating each of them (Carroll et al. 1998).

The rationale would also be a logical place to capture the source of the requirement. Knowing which stakeholders, and which stakeholder category they fit into would be useful if questions arise about the requirement that require clarification. Knowing the requirement source would assist in the prioritization of the requirement by identifying the interested parties. Rationale can also associate requirements identified and refined during the requirements engineering process with the original customer requirements and provide "rich traceability" (Dick 2005; Hull et al. 2002). The rationale would also provide the intent behind the requirement, or the stakeholder goal(s) that the requirement addresses. The mapping of goals to requirements can be used later to determine which requirements would require adjustment if the goals change later in the development process. The rationale would also be a place where dependencies or conflicts between requirements can be identified. Knowing the source and intent of each conflict will be useful when determining the best way to resolve conflict and inconsistency.

11.2.2 Achieving Consensus

An important part of the RE process is the negotiation that needs to take place between the various stakeholders. The different groups approach the system from different viewpoints and may have conflicting goals. The rationale for the requirements is a key element in the negotiation process by providing a means for identifying conflicts and explicitly stating the arguments of all participants. The collection process itself was found to be useful during field trials using itIBIS and gIBIS (the textual and graphical versions of the Issue Based Information Systems approach) (Conklin and Burgess-Yakemovic 1995). Structured rationale capture assisted with team communication by making meetings more productive. The Compendium approach (Shum et al. 2006) is an IBIS-based collaboration support system that is used to capture stakeholder needs via a "dialogue map" that aids in collaboration by structuring discussion and capturing the "meanings and ideas" of the group.

The role of rationale in requirements negotiation is a key element in the WinWin approach to requirements negotiation (Boehm and Kitapci 2006; Boehm et al. 1994). An ontology defining the rationale in WinWin was developed by Bose (1995) and describes what the attributes are for the WinWin rationale elements (Winconditions, Options, Issues, and Agreements). The goal of the WinWin approach is to make "winners" of the system's stakeholders. The EasyWinWin tool assists in group facilitation to aid in determining what the win conditions are, prioritizing the win conditions, identifying what the issues are, and capturing the decision rationale (Grünbacher and Boehm 2001). Experiments performed using students demonstrated that the WinWin approach assisted with distributed collaboration, aided in cooperation, reduced friction between team mates and helped the students to focus on the key issues (Boehm and Egyed 1998). There are more than one hundred real-world projects that have used EasyWinWin (Boehm and Kitapci 2006).

An alternative approach to requirements negotiation and validation is the Software Quality Function Deployment (SQFD) approach (Ramires et al. 2005). SQFD builds a matrix that gives correlation values between specifications and requirements where the stakeholders provide the correlation values. The MEG groupware tool was built to support SQFD and added rationale, in an adapted IBIS format, to the SQFD matrix. The IBIS component captures stakeholder positions and arguments. The evaluation of the requirements is achieved using a majority voting scheme where votes are weighted depending on how each stakeholder participated in past decisions (Win-Win, Win-Lose, or Lose-Lose).

11.2.3 Requirements Inconsistency

Since requirements are obtained from a variety of stakeholders and sources, there is a risk that inconsistencies may arise. It is important to identify inconsistencies so they can be handled appropriately, whether through resolution, avoidance, deterring, or ignoring (Nuseibeh et al. 2000). There are a number of approaches to performing consistency checking in requirements. The C-Re-CS system (Klein 1997) captures requirements and their rationale in a semantic net structure. The system contains exception management services that check for completeness, correctness, and consistency in the requirements; identify problem diagnosis using a knowledge base of general requirements problems; and identify potential resolutions to the problems based on past knowledge of general problems. The knowledge base is structured as a taxonomy of diagnoses from more general to more specific that is traversed similarly to a decision tree based on questions and answers.

Reiss (Reiss 2002) has developed a constraint-based, semi-automatic maintenance support system that works on the abstracted code, code, design artifacts, or meta-data to assist with maintaining consistency between artifacts. The CLIME software development environment checks for consistency between UML class diagrams and source code; between UML interaction diagrams and code; test cases to source code (to ensure unit tests have been run if a method was modified); documentation to source code; source code to documentation; and also checks code and documentation to ensure that certain pre-set standards (such as naming conventions) are followed (Reiss et al. 2003).

11.2.4 Requirements Prioritization

Recent work on Value-Based Software Engineering has begun to address the problem that software development efforts treat each requirement (and other development artifacts) as if they were of equal value (Boehm 2006). In reality, some requirements are more important to the stakeholders than others. When decisions need to be made to decide what requirements should be implemented first or should be given the most resources, it would make sense to base these decisions on the relative value of the requirements and prioritize them.

Karlsson and Ryan (1997) propose using a cost-value approach to prioritize software requirements. This method uses the Analytic Hierarchy Process (AHP) (Saaty 1980) to perform pairwise comparisons of the requirements. Customers and users use AHP to provide relative value and software engineers use AHP to provide relative cost. This was an effective method for determining priorities but does have scalability issues for large numbers of requirements.

Rationale can also be used to assist with the requirements prioritization process. The rationale behind each requirement can capture the underlying intent behind the requirement. The Software Engineering Using RATionale (SEURAT) tool (Burge and Brown 2006) allows the rationale for each requirement to be captured. This rationale can serve as a basis for negotiating requirement importance and could potentially be used to compute rankings for the alternatives.

11.3 Requirements Traceability

Requirements traceability typically refers to the ability to trace from the requirements all through the development process. The goal of traceability is to ensure that all system requirements are met. Requirements traceability is a key element in requirements management and is required to assess the impact and consequences of requirements changes (Nuseibeh and Easterbrook 2000).

Requirements can be traced in two directions. Tracing a requirement backwards refers to tracing back from the requirements specification to the origins of the requirement. Tracing a requirement forwards traces from specification through implementation and test. These two directions are referred to as Pre-RS traceability and Post-RS traceability, respectively (Gotel and Finkelstein 1994). The rationale for the requirements and for the developed system can aid in both kinds of traceability.

Pre-specification traceability is one of the more neglected forms of traceability (Gotel and Finkelstein 1994). The ability to know the origins of a requirement can be used later on if the requirement needs further clarification. Unfortunately, this information is often difficult to obtain. One way to track the origin of a requirement would be through the rationale for the requirement. The rationale would provide information on who argued for (or against) its inclusion and what the reasons behind the choice were. This information can be very useful in future development if the requirements need to change.

Post-specification traceability is what most developers think about when they think about requirements traceability – the ability to trace from the requirements through to the test cases in order to ensure that the software system meets its specification. Rationale can assist with post-specification traceability. The requirements, both functional and non-functional, can appear in the arguments for and against the many decisions made when designing and implementing the software. Each alternative chosen would eventually map to some development artifact, whether a section of a document, elements in a UML diagram, or the code itself. The arguments for choosing that alternative consist of requirements and non-functional requirements. The mapping from the alternative to its implementation would then provide traceability to those requirements. An example of this is the SEURAT system (Burge and Brown 2006) which captures traceability between code elements and alternatives.

The "rich traceability" proposed by Hull et al. (2002) supports both pre- and post-specification traceability by representing requirements at different levels—stakeholder requirements, system requirements, and design requirements. Rich traceability contains "satisfaction arguments" that can be supported by domain knowledge as well as information from other sources such as the output of modeling tools. These satisfaction arguments indicate how requirements relate to each other, in particular they capture when all of the requirements at one level are necessary to satisfy requirements at the level above (conjunction) or if any one requirement is needed (disjunction). For example, it may be necessary that all of a set of system requirements must be satisfied to satisfy the stakeholder requirement they relate to or it may only be necessary that one be satisfied.

Non-functional requirement (NFR) traceability is also important. The relationship between rationale and NFRs is described in the following section. Surveys of NFR traceability approaches can also be found in Hayes et al. (2005) and Cleland-Huang et al. (2005).

11.4 Rationale and Non-Functional Requirements

While functional requirements describe the function of a system or device, non-functional requirements describe how the system or device should accomplish that function given "the constraints of a non-ideal world" (Thayer & Dorfman 1990). Non functional requirements often refer to software quality and are related to Boehm et al.'s "Quality Characteristics." (Boehm et al. 1979). Roman (1985) describes NFRs as restricting the types of solutions under consideration. NFRs are not directly related to specific system components and often involve aggregate system behavior (Manola 1999). Research involving non-functional requirements and their impact on software development is taking place in a number of areas, many of which fall into the category of "separation of concerns" (Workshop 2000; Ossher and Tarr 1999). Concerns can fall into many, often overlapping, categories and can describe concerns about features, requirements, extensibility, performance, and reliability. Many categories of concerns have been proposed but the common thread is that each category describes attributes of a system that "cross-cut" the system's structure and/ or functionality.

Functional requirements describe the functionality that a system needs to provide in order to satisfy the needs of its stakeholders. Non-functional requirements describe how well the system needs to perform that functionality or, in some cases, how the development effort needs to proceed in order to meet the needs of the customer and the developing organization. One way that the NFRs can be captured during requirements engineering and all through development is in the rationale for the system. The NFRs would appear as arguments for and against different alternatives considered. The rationale can be analyzed to assess the impact of various NFRs on the software product and to determine how the decisions made might change if NFR priorities change.

11.4.1 Non-Functional Requirement Categorization

When working with NFRs, it is often useful to work with a set vocabulary, or ontology, of terms. In rationale-based systems, a common vocabulary of keywords is needed to support semantic inference (Burge and Brown 2000). There are several different ways that NFRs have been organized or grouped. Bruegge and Dutoit (2000) referred to NFRs as "design goals" and broke them down into five groups: performance, dependability, cost, maintenance, and end user criteria. Chung et al. (2000) provides an unordered list of NFRs and also hierarchies of NFRs for performance and auditing.

Some categorizations emphasize NFRs that relate to software quality and have formed quality measure hierarchies. The ISO/IEC 9126 software product quality standards (Jung et al. 2004) give six characteristics (functionality, reliability, usability, efficiency, maintainability, and portability) as well as 27 sub-characteristics. The CMU Quality Measures Taxonomy (CMU 2002) organizes quality measures into Needs Satisfaction Measures, Performance Measures, Maintenance Measures, Adaptive Measures, and Organizational Measures.

11.4.2 The NFR Framework

The view that quality characteristics are important when developing a software system was the driving force behind development of the NFR Framework (Chung and Nixon 1995). The NFR Framework uses non-functional requirements, represented as Softgoals, to drive the software design process (Chung et al. 2000). This process produces the design, because the process is driven by the NFRs—its rationale. The NFRs are represented in a softgoal interdependency graph. The graph allows traceability from requirements to design decisions and from design decisions back to the requirements considered (Chung and Yu 1998). If requirements are changed, the goal graph can capture a historical record that relates new requirements to the old ones (Chung et al. 1996).

Cysneiros and Leite (2004, 2001) focused on how NFRs could be incorporated into the conceptual models represented in UML. They chose to create two views of the system: a NFR view, built on the NFR Framework (Chung et al. 2000) and a functional view, captured in UML. These two views should be connected at "convergence points." A Language Extended Lexicon (LEL) was built to contain the vocabulary used for the functional requirements and links to the NFRs. The LEL is generated first and is used in constructing the functional and non-functional views.

The NFR Framework was also used to support the Goal Centric Traceability (Cleland-Huang et al. 2005) approach. Goal Centric Traceability consists of four phases: goal modeling, impact detection, goal analysis, and decision making. The goal modeling phase uses Chung's Softgoal interdependency graph (SIG) (Chung et al. 2000) to capture the NFRs and trade-offs. Impact detection automatically creates links between the SIG elements and a functional model of the system captured in UML class diagrams using ontological keywords. Goal analysis propagates changes made to the goal contributions by the user through the SIG to determine their impact.

11.4.3 SEURAT Argument Ontology and NFR Prioritization

The ability to inference over the rationale has many different uses. One use, demonstrated in the SEURAT system (Burge and Brown 2006), is to evaluate the impact of changing priorities over the life-time of a system. This capability was supported by the use of an Argument Ontology (Burge 2005). This ontology, based on the NFR taxonomies described earlier (Bruegge and Dutoit 2000; Chung et al. 2000; CMU 2002; Jung et al. 2004) and extended to incorporate additional criteria, contains a hierarchy of reasons for making software decisions. The base elements of this ontology are Affordability Criteria, Adaptability Criteria, Dependability Criteria, End User Criteria, Needs Satisfaction Criteria, Maintainability Criteria, and Performance Criteria. The hierarchy then subdivides these items into more detailed criteria (up to four levels deep). The Argument Ontology contains 277 terms and is documented in (Burge 2005).

SEURAT uses the rationale to re-evaluate the support for each decision whenever the importance (priority) of an element in the argument ontology changes. This can show which (and how many) alternatives may need to be reconsidered. Another use of rationale supported by SEURAT is to detect relationships between functional requirements and the NFRs in the Argument Ontology. This can be done by looking for the ontology entries that appear in arguments along with the functional requirements. This may indicate a relationship between the goals depicted in the ontology and the functional requirements.

The rationale can also be used to analyze the reasons for and against the decisions made in order to determine how, and by how much, the goals determined during the requirements engineering process ended up influencing the final system. This is something that can be done during development to ensure that the program is staying on-track and that the decisions are made in accordance with customer priorities and also after development to learn what might be the important factors to consider when developing future systems.

11.4.4 NFRs and Conflict Representation and Detection

The relationships between NFRs and the relationships between NFRs and FRs can be used to identify conflicts between requirements. In the case of NFR to NFR conflicts, it is important to determine how these requirements interact to avoid situations where one is met at the expense of another. This need to achieve "balance of attribute satisfaction" was the impetus behind the Quality Attribute Risk and Conflict Consultant (QARCC) (Boehm and In

1996). Given a win condition generated using WinWin, QARCC uses a knowledge base of architecture (product) and process strategies for achieving quality attributes to check for conflicts. The knowledge base identifies the positive or negative impact that an architecture strategy has on affected quality attributes. The quality attributes are stored in a hierarchy where attributes at the highest level of abstraction, the "primary quality attributes" are mapped to stakeholder roles. Conflict detection is supported by the rationale captured in SEURAT by storing tradeoffs between quality attributes as background knowledge that is then used to detect conflicts. This differs from the approach used in QARCC by capturing the tradeoffs directly and not relative to a specific architectural decision.

Egyed and Grünbacher (2004) use the quality attributes to detect conflicts in functional requirements. In their approach, functional requirements have requirement attributes that relate to qualities such as efficiency, usability, security, etc. A cooperation and conflict model gives the relationships between qualities (positive, negative, or no effect). If a requirement has quality attributes that conflict with those of another requirement, that may indicate a conflict between the two requirements. This approach in and of itself would likely generate numerous false positives so it is augmented with trace analysis to only report conflicts between requirements that effect the same part of the code. The traces are generated by running the test scenarios that test each requirement.

11.5 Goal-Based Requirements Engineering

Requirements engineering can be viewed as the process of transforming stakeholder needs, or goals, into requirements that describe the system that will meet those names or goals. Even in cases where the stakeholders explicitly express their requirements, the system may be more successful if the goals behind those requirements can be expressed so that alternative ways to meet those goals can be explored (Antón and Potts 1999). In this section, we will look at two approaches involving goals: Goal-Based Requirements Analysis (GBRAM) (Antón and Potts 1998) and Goal Oriented Requirements Engineering (GORE) (van Lamsweerde 2001).

11.5.1 Goal-Based Requirements Analysis

In GBRAM (Antón and Potts 1998), goals are defined in two phases: goal analysis and goal refinement. In goal analysis, the analyst explores various information sources to identify possible goals and classify them according to goal dependencies. In goal refinement, the goal set is pruned if necessary, goals are analyzed to identify obstacles towards the goals, and goals are operationalized (turned into formal requirements).

Specifications and scenarios (sometimes in the form of Use Cases (Antón et al. 2000)) are used as in puts to the goal identification process. One method used to identify goals is to look for verbs such as "avoid" or "improve" that are then followed by a desirable or undesirable condition. These verbs are also uses to categorize goals into categories based on the verb used such as. This categorization is used to separate user goals from system goals. User goals are identified as "achieve" goals while system goals (how the system responds to the user goals) are identified as "make" goals. The categorization differentiates between providing capability and providing information by using "notify" and "inform" to describe providing information and using "provide" and "allow" to describe providing capability (Antón et al. 2000). The goal categorizations used vary depending on the domain. The CommerceNet Web Server project described in (Antón and Potts 1997) used avoid, ensure, improve, increase, keep, know, maintain, make, and reduce while the e-commerce system analyzed in (Antón et al. 2000) used allow, achieve, make, provide, inform, ensure, and notify as the goal categories.

11.5.2 Goal-Oriented Requirements Engineering

Goal-oriented requirements engineering (GORE) focuses on the use of goals to drive requirements engineering (van Lamsweerde 2001) (van Lamsweerde 2004). Goals can be functional goals, which are then used to build use cases and other "operational models" or quality goals that describe "preferred behavior" and are used to compare different alternatives as well as posing constraints (van Lamdsweerde 2004). The level of abstraction can also vary from high level goals that are strategic to low level goals that describe technical concerns (van Lamsweerde 2001).

The KAOS (Keep All Objectives Satisfied) method (van Lamsweerde and Letier 2000) represents goals and obstacles (undesirable conditions) in a formal temporal logic. The KAOS construct specification consists of two levels: a semantic net layer declaring the concept and its relationship with other concepts and a formal assertion layer that gives a formal definition. The second, formal, layer is optional and is used for formal reasoning while the semantic net layer supports modeling, traceability, and re-use. The goal specification defines the goal, and the property it should hold (achieve, cease, maintain, avoid), what other objects are involved, the parent goal, sub-goals that it should be refined to, and an informal description of the goal. The specification also contains the formal layer expressed in temporal logic. While goals describe desired behaviors, obstacles describe undesirable behaviors. Obstacles can be broken into five types: non-satisfaction obstacles that keep goals from being satisfied; non-information obstacles that obstruct information dissemination; inaccuracy obstacles that obstruct object state consistency; hazard obstacles that interfere with safety goals; and threat obstacles that interfere with threat goals.

van Lamsweerde and Letier (2000) define a requirements elaboration method that elaborates and operationalizes goals while also defining obstructions to those goals. This process starts with elaboration, where goals are refined; the object capture that determines what objects are involved (objects can be entities, relationships, or events); operation capture that finds object state transitions; operationalization, which determines pre and post conditions; and finally responsibility assignments to identify alternative assignments and select alternatives based on non-functional goals. Obstacles and alternative resolutions are identified during the elaboration phase.

The GORE process is also supported by a software environment, GRAIL (Goal-Driven Requirements Analysis, Integration, and Layout), which supports editing, semantic checking, and views (Darimont et al. 1997). GRAIL contains a text editor for requirements acquisition and to check syntax and semantics. GRAIL can also present a graphical view of the specification.

11.5.3 Relationship to Rationale

These methodologies both utilize rationale. In GBRAM (Antón and Potts 1998), the rationale for requirements provided by the stakeholder is used during the refinement process to determine if there are additional requirements that need to be generated. As the refinement process proceeds, the rationale is tracked so that any unresolved issues can be monitored and eventually resolved. Each requirement generated in the GBRAM process is annotated with rationale: the questions, answers, alternatives, and scenarios that were generated and used during refinement.

Defining requirements using the GORE method produces the rationale for the requirements in the form of the goals that they were derived from. The goal hierarchy that resulted in the final requirement definition can be traced back to determine the rationale.

There are a number of places within the methodology where decisions need to be made. One is in the assignment of responsibility for the terminal goals to "agents": entities (humans, programs, devices, etc.) that perform operations or agents that monitor an object. Assumptions are defined as terminal goals that are assigned to "agents in the environment", while requirements are defined as terminal goals that are assigned to "agents in the software" (van Lamsweerde and Leiter 2000). The alternatives are captured in the GORE process and the selection criteria can be captured as well.

There are also alternative resolutions to obstacles defined during the goal elaboration phase. The resolution strategies range from obstacle elimination to obstacle tolerance (van Lamsweerde and Leiter 2000). The choice of resolution strategy depends on the likelihood and severity of the obstacle. The alternative resolutions and reasons for resolution selection should be documented in the rationale.

11.6 Adapting to Changing Requirements

As stated earlier, failure to manage requirements, or more specifically manage requirements change, is a major cause of project failure. Managing requirements change requires addressing the following issues: identifying (the need for) change, impact analysis, determining when changes conflict, negotiation, prioritizing changes, change measurement, risk assessment, change estimation, planning (scheduling), and change learning (Lam et al. 1999).

These issues are strongly related to each other. For example, negotiation is heavily involved when determining the need for change, prioritizing changes, and scheduling change. Rationale has been shown to be an effective strategy in supporting negotiation by allowing the views of all the participants to be captured in a formal or semi-formal manner. The ability to use the rationale in evaluating alternatives can be helpful in prioritization as well, especially if the rationale captures the importance of different evaluation criteria.

Impact analysis and risk analysis are closely related. A requirement having a higher impact on the system will bring a higher risk. If the requirement is a change to an existing one, the rationale could be used to determine what parts of the system were affected by the original requirement so that those could be modified. This is supported by systems such as SEURAT (Burge and Brown 2006) which use requirements as part of the argumentation and map the selected alternatives to the code that implement them. The ability to perform impact assessment also assists with change measurement since the impact on the system is related to the amount of change needed. The impact assessment results will assist with change estimation as well.

One issue that rationale is especially helpful with is in supporting the process of determining when changes conflict. The rationale records the intent behind the current choices made in a system and should also capture what tradeoffs were made. New requirements can be assessed against known tradeoffs. Another use of rationale is to avoid repeating mistakes that occurred in the past. If a change is proposed that was rejected earlier, the rationale will capture that decision and inform the analyst that there is a potential problem.

The goal of "change learning" is to collect information about changes that have occurred so that when similar changes happen in the future, information about that change will "reduce surprise" (Lam et al. 1999). Change information for a change, or type of change, can be captured in its rationale. The rationale would provide the reasons for the change, how it was made, and other pertinent information such as cost. Rationale also helps with learning by providing the intent behind the original requirements.

11.7 Summary and Conclusions

Requirements Engineering is a crucial component of all software developments. The ability to successfully capture stakeholder needs and represent them in a way that they can then be used to drive software development had a significant impact on the success of the software project.

In this chapter, we describe the requirements engineering process and some key aspects including requirements elicitation, negotiation, prioritization, and traceability. We also discuss research in non-functional requirements and goal-based requirements engineering. These areas have strong ties to rationale. By using NFRs to drive system design, the NFR framework captures the rationale for each decision. Goal-based requirements engineering examines the goals that drive each requirement, i.e. its rationale.

Because of the criticality of requirements, and the high costs incurred if requirements are incorrect, incomplete, or mismanaged, capturing the rationale for the requirements should be a necessary step in the RE process.

12 Rationale and Software Design

More has been written about software design rationale than about any other topic in research on software engineering rationale. Much work has gone into identifying the value of design rationale for software developers, maintainers and users; but realizing this value requires that approaches to rationale capture and delivery be successfully integrated into the processes of software design. This chapter looks at the complexities of this task and a variety of approaches that researchers have adopted for dealing with them.

12.1 Introduction

A crucial goal of Rationale-Based Software Engineering is to effectively capture and use rationale throughout software design. Both capture and use of design rationale present problems for researchers and practitioners, though the challenges of rationale capture are by far the more challenging. To solve these problems, it is crucial to understand how processes of rationale capture and use relate to what software designers do. More specifically, it necessary to understand how decision-centric and usage-centric approaches to rationale fit into, or fail to fit into, the processes of software design.

The processes that practicing software designers use are varied. Some are the product of their personal experiences and beliefs. Some are prescribed by design methods that they subscribe to. Some are dictated by the SE tools that designers use. The variety of processes in use is likely to continue increasing as methods and tools evolve over the coming decades.

Given the variety of design processes and rationale approaches, the question arises as to how to go about discussing the fit of rationale approaches to design. A comparison of all rationale approaches with all design processes is clearly beyond the scope of this chapter. Instead, the chapter will look for underlying principles of fit and misfit.

12.1.1 The Nature and Importance of Software Design Rationale

12.1.1.1 The Nature of Software Design Rationale

Software design rationale (SDR) is the reasoning used in making decisions about the design of software. Most of the literature on design rationale deals exclusively with the elicitation and structuring of rationale from designers. But to understand fully the issues of SDR, it is important to recognize that not all the rationale used by the designers in a given software project is generated by those designers. Some of this rationale comes from stakeholders involved in SE activities other than design and includes information about requirements as well as feedback from construction and use of prototypes and earlier versions of the software. This externally generated rationale can also include information about the rationale for and outcomes of earlier projects.

12.1.1.2 The Importance of Software Design Rationale

Support for the capture and use of rationale generated by designers is important because it can improve design and other SE activities, such as construction, maintenance (Burge and Brown 2006) and the management of software projects. It can also facilitate coordination and collaboration in development teams as well as participation by users in development. The argument has also been made that it can aid the users of software in understanding complex, high-functionality systems (Haynes 2006).

Also important, however, is the design rationale that is not generated by a project's designers. Systematic use of such externally generated rationale provides an *intelligence augmentation* (IA) strategy, i.e. a way of augmenting the rationale of designers to enhance the quality of their design efforts. Externally generated rationale includes feedback from construction and use, which is one of the driving forces behind iterative approaches to software design and development. Enriching this feedback and other external sources of rationale might be the most promising means for helping designers to cope with the increasingly pressing problems of software development described in Chapter 1.

12.1.2 Objectives of this Chapter

The main objective of this chapter will be to identify underlying principles of fit and misfit between rationale approaches and design processes. It will do this by means of two kinds of analysis. The first is a general, theoretical analysis of rationale approaches and design processes. The second is an analysis of concrete examples of attempts to integrate rationale approach into software design processes.

Section 12.2 looks in a general, theoretical way at the issues of relating rationale approaches to software design processes. Section 12.2.1 looks at the ways in which decision-centric and usage-centric approaches to rationale fit into design. Section 12.2.2 deals with ways in which prescriptive and descriptive roles of rationale approaches can support and conflict with design. Section 12.2.3 examines the roles of rationale for design space analysis and deeper reflection relate to each other and to the design process.

Section 12.3 looks at specific approaches to tailoring rationale approaches to software design. Section 12.3.1 surveys a variety of approaches that researchers have devised for integrating rationale into the design of software architecture. Section 12.3.2 then speculates on what this highly diverse research suggests in the way of principles for fitting rationale processes into the processes of design software architecture.

Finally, Section 12.4 summarizes the chapter and draws conclusions about the state of research on software design rationale.

12.2 Relating Rationale Approaches to Software Design Processes

In determining how rationale approaches relate to design processes, we can make good use of three basic distinctions. One is between decision-centric and usage-centric rationale. A second is between the descriptive and prescriptive roles of rationale approaches. A third is between the rationale for design space analysis and deeper reflection. These distinctions reveal important information about the compatibility of rationale approaches and design processes.

12.2.1 Decision-centric and Usage-centric Rationale Approaches

Decision making is the concept that most obviously connects SDR to software design processes. This section with therefore start by looking at the decision-centric approaches, which explicitly represent decision making processes. The results of this analysis will then be used to analyze the usage-centric rationale approaches, which do not represent such processes.

12.2.1.1 Decision-centric Rationale and Design Processes

What makes it possible to see the fundamental connection between decision-centric rationale processes and design processes is the fact that both
deal with *decision making*. In particular, both deal with *decision tasks* and
the *evaluation of proposed decisions* as a way of arriving at decisions.
Typically, both also deal with *decision alternatives*, i.e. multiple, alternative, proposed decisions. These things might be represented differently in a
given rationale approach and a given design process used by a software
designer; nevertheless, understanding their equivalence makes it possible
to see the crucial similarities and differences between the rationale approach and the design process.

The next crucial similarity is that each decision-centric rationale approach
and each design process necessarily has *a way of evaluating proposed decisions*, i.e. decision alternatives. Finally, we can see that each decision
task can, and usually does, have *a decision* to adopt one of the proposed
decision alternatives.

If we look at the above-stated model of decision making we can see that
practically everything a software designer does is part of some decision
making process of this type. We can also see that every decision-centric
approach to rationale models decision making in this way. As a consequence, the question of how and where such approaches to rationale fit into
design appears simple to answer: they fit everywhere into design processes and they fit well.

The problem is that this notion of an extensive and deep fit between
decision-centric rationale and design process is hard to reconcile with the
reality of the rationale capture problem. The fact that this problem is currently the greatest obstacle to use of rationale approaches in software projects seems to suggest a fundamental *misfit* of some sort between rationale
approaches and design process. This section will attempt to understand this
dilemma by looking for ways in which rationale approaches can come into
conflict with design processes. But first it will look at how usage-centric
rationale fits into the overall design process.

12.2.1.2 Usage-centric Rationale and Software Design Processes

According to the definition given above, something can count as SDR only
if it plays a role in helping to make design decisions. This might seem to
suggest that only decision-centric rationale approaches deal with SDR, but
this is not the case. A crucially important use of usage-centric rationale approaches—such as Scenario-Claims Analysis (SCA) (Carroll and

Rosson 1996)—is to provide rationale that informs the decision making of designers.

The contribution of usage-centric rationale approaches like SCA is to recognize that organizing rationale around decisions is not the best way to elicit and characterize some of the rationale needed for making appropriate design decisions. A design must in large part be judged in terms of its consequences for its users, and the best way to identify these consequences is to document the evaluation of system features by users as they interact with the system. This information can then be fed back to the system designers in the form of argumentation rationale that prompts them to revise their decisions about the design of the system. Thus, usage-centric rationale, such as that produced by SCA, gets its value for design by informing design decision making, but it does so by providing feedback that gets designers to *change their previous decisions*. SCA is thus part of an iterative design process that Carroll and Rosson have labeled "the task-artifact cycle." So ultimately, a complete account of SDR must show how usage-centric rationale becomes part of the evaluation of decision alternatives in decision-centric rationale for software design.

12.2.2 Prescriptive and Descriptive Roles of Rationale Approaches

The distinction between the prescriptive and descriptive roles of rationale approaches, explained in Chapter 1, reveals various ways in which these approaches can be compatible or incompatible with design processes. Rationale approaches can play various descriptive and prescriptive roles in design, and these roles intrude into the design process in different ways and to different degrees. The intrusiveness of rationale approaches is an especially important topic because it is at the center of a controversy amongst rationale researchers about the difficulties of getting rationale approaches used in practice. The focus of this controversy is the *rationale capture problem*, which is widely regarded as the main obstacle to practical application of rationale. One side of this controversy advocates the use of traditional approaches to capture, which tend to be relatively intrusive. The other side, which has emerged over the past decade or so, argues that the intrusiveness of traditional approaches to capture is the main cause of the capture problem. Making sense of this controversy requires a detailed understanding of the varieties of prescriptive and descriptive roles of rationale and the ways in which they intrude into design. This section starts by looking at the prescriptive roles and then looks at the descriptive.

But before describing the intrusiveness of rationale approaches it is important to state some words of caution. It would be a mistake to regard intrusion into design as necessarily bad. There is no way to improve design without altering it, and this means intruding into it. Even if one believes that the intrusiveness of rationale methods has been the central barrier to rationale capture, it would be foolish to conclude that designers are against all intrusions into design. After all, this would imply that they were opposed to the idea that design could be improved.

12.2.2.1 Prescriptive Rationale Approaches and Design Processes

Prescriptive approaches to design rationale attempt to alter the thinking of designers in order to improve design. While the intention of such approaches is to be useful to designers, there exists the possibility that they will interfere with the way designers prefer to work. In fact, one possible explanation for the difficulties in getting designers to adopt rationale approaches is that they do not like being told how to do their jobs. But before accepting such a glib explanation, it is prudent to look more closely at the varieties of prescriptive roles and how they affect design.

Two ways rationale approaches can be prescriptive. Rationale approaches can be prescriptive by *informing design*, i.e. by providing information for designers to think about in making decisions; or they can be prescriptive by *prescribing processes* for designers to follow in making decisions. An example of the former is the approach of Fischer et al. (1996) (described in Chapter 4 of this book) which uses knowledge-based critics to supply designers with information from a collection of rationale structured using the PHI variant of IBIS. An example of the latter is the process-oriented approach of IBIS by Conklin and Burgess-Yakemovic (1996).

Intrusiveness of these two ways of being prescriptive. The approach of *informing design* represents a relatively minor intrusion into the design process. Thus, in the work of Fischer et al. the design process is only intruded into briefly and intermittently when critics detect violations of rules-of-thumb for design and display rationale to help designers determine whether it makes more sense to follow the rules or break them.

By contrast, the approach of *prescribing processes* is a much greater intrusion on the design process. Thus, in the work by Conklin and Burgess-Yakemovic, designers are restricted to what these authors call "'legal moves' in the IBIS design conversation" (Conklin and Burgess-Yakemovic 1996) throughout the design process. They describe the use of

IBIS to structure meetings, but IBIS can also be used to structure the work of individual designers.

It is clear that, in principle, prescriptive approaches to rationale should be able to justify their intrusiveness by their benefits to designers. All prescriptive uses of rationale in design are, by definition, aimed at aiding designers. The question seems to be what types and degrees of intrusions designers are likely to regard as worthwhile.

Conklin and Burgess-Yakemovic as well as others, e.g. Shum et al. (2006), have reported many cases where designers found the use of process-prescribing rationale methods produced higher quality design. Despite such reports, it is clear that effective rationale capture remains a largely unsolved problem. So it is important to see if there are additional ways in which rationale might conflict unacceptably with processes that designers choose to use.

Ways in which designers might view intrusiveness as bad. There are two respects in which a prescriptive rationale approach can dictate how decisions are made. The first is that it can require use of a conceptual schema for categorizing and inter-relating the rationale used in decision making. For example, when IBIS is used in a prescriptive approach, its schema requires that rationale be stated in the form of *issues*, *positions*, *arguments* and *resolutions* and that these elements be linked together only in certain ways using a given set of relationships. Other rationale schemas have similar requirements when used in this prescriptive manner.

There are a number of reasons that designers might have for viewing schema-based approaches to decision making as undesirable. They might feel that a given schema does not fit their individual, and perhaps highly skilled, modes of reasoning about design. Or they might be committed to using a software design method or tool that does not allow rationale to be organized according to the given schema.

Much of the rationale literature has been devoted to devising new schemas, and not infrequently this work is based on a claim that the difficulties of rationale capture derive from mismatches between schemas previously used and the way in which designers naturally organize their thoughts. Ironically, there is little or no evidence that changing the schema has resulted in more effective capture of rationale. This suggests that the problem might not be any particular schema, but the use of schemas in general.

Schön's theory of Reflective Practice suggests that use of a rationale schema fundamentally conflicts with some of the cognitive processes required for design. Schön argues that designers alternate between two complementary types of processes: an intuitive process of action and a rational process of reflection. A cornerstone of his theory is that a designer can only engage in one of these processes at a time. Using a rationale schema to structure design thinking is a kind of rationale reflection. Thus, the attempt to use schema-based rationale throughout the design process is in effect an attempt to turn all of design into rationale reflection. According to Schön this makes design impossible. It should be noted that Marshall and Shipman have made a similar but more general argument about the counterproductive nature of schemas for human-computer interaction (Shipman and Marshall 1999).

Intrusiveness of elicitation and structuring procedures. There is a second respect in which a prescriptive approach to rationale can dictate the way in which decisions are made, and that is by prescribing the procedures used for eliciting and structuring rationale. These procedures can also conflict with the processes that designer prefer to use, either because they prefer to think in a given way or because they are committed to using software design methods or tools that conflict with the elicitation and structuring procedures mandated by the prescriptive approach to rationale.

Temporal intrusiveness. Wherever a prescriptive rationale approach does not conflict with the processes designer prefer to use, there might seem to be no obstacle to using the approach in conjunction with the preferred design processes; but there is one more problem that is potentially a "show stopper." The problem is that using any rationale that involves having the designer document rationale or even participate in the documentation of rationale is likely to be very time consuming. There are two questions that need to be answered before the documentation of rationale can be justified. One is whether the designers have enough time to participate in the documentation. The other is whether having the designers spend time on this documentation is of greater value than having them spend that time on design itself. The latter question indicates that it is not enough to consider the absolute cost of documenting rationale; it is necessary to consider the *lost opportunity costs* of such documentation. It seems likely that the inability to answer these two questions in the affirmative has played a large role in limiting the capture of rationale in real-world software projects.

12.2.2.2 Descriptive Rationale Approaches and Design Processes

In retrospect, it might seem obvious that procedurally prescriptive rationale approaches are inherently intrusive on design processes; and this might create the expectation that when playing a purely descriptive role rationale approaches would necessarily be less intrusive. Things are not that simple, however, because designers often do not state their rationale, much less document it, when they design. Obtaining a detailed record of their rationale might therefore require a concerted effort to elicit it from them and to record it. This effort has the potential for being highly intrusive on the design process. As a consequence, researchers who feel that intrusiveness is the root of the capture problem have sought various ways of describing designers' rationale in ways that are less intrusive.

Intrusiveness of various approaches to describing rationale. One highly intrusive approach to eliciting rationale is to systematically interrogate designers about what decisions they make and the reasoning for each decision. This approach becomes maximally intrusive when it uses a rationale schema and dictates the order in which statements are elicited. Such an approach is nearly as intrusive as the most aggressive procedurally prescriptive approaches but has the further disadvantage of offering no obvious payoff to designers as a motivation for them to tolerate the intrusion.

The approach of systematic interrogation can be done either in process-oriented or structure-oriented mode, i.e. either to produce a history of the design process or a "logically" structured record of design rationale without any indication of the process by which it was produced. In the former case it intrudes on every step of the design process. In the latter case it can be done in retrospect and so could be intrusive only in the sense of requiring designers' time. Of course, there may be some question about how accurate such retrospective accounts of rationale are likely to be.

Perhaps the most extreme example of intrusive elicitation of rationale using a purely descriptive approach was found in the use of PROTOCOL, the first software designed exclusively as a rationale management system (McCall 1979). This system used a systematic interrogation approach based on the PHI schema and had a fixed order in which statement types were elicited. The approach was applied recursively in the sense that each response became the subject of further interrogation, in a manner somewhat reminiscent of the Eliza system (Weizenbaum 1966). A rationale elicitation session only ended when users were unable or unwilling to give further responses. While the system was highly effective in eliciting large quantities of rationale, users generally found the experience extremely tiring and few were willing to repeat it. Someone observing this effect commented that the system had given a whole new meaning to "exhaustive enumeration" (McCall 1979).

The QOC approach is much less intrusive than the approaches described above. It is exclusively targeted at *describing* the rationale for design—usually software design—and employs a structure-oriented approach. It uses a schema in eliciting rationale, but apparently does not dictate the precise order in which statement types are stated. Most important of all is the fact that QOC does not intrude into the design process directly, because it is not used while design is taking place.

QOC can, however, *indirectly* intrude on design to various degrees, depending on how it is used. The authors of QOC insist that that the rationale for a project should itself be designed, so the crucial question is how much the designers themselves need to be involved in this process. The more designers are involved in designing the design rationale, the more time will be required of them and, consequently, the more temporally intrusive QOC will be on design. Users of the QOC approach, however, might minimize this intrusiveness by eliciting only the raw rationale from the designers and employing other people, e.g. rationale experts, to refine and design this rationale.

Descriptive approaches lend themselves to highly unobtrusive capture, including both automated capture and automated structuring of rationale. The work of Myers et al. (1999) (described in Chapter 4) uses a completely non-intrusive approach that captures rationale by recording the actions of designers using a CAD system. Schneider (2006) does a similar kind of non-intrusive capture of rationale from software engineers as a by-product of their use of development tools. McCall and Mistrik (2005) use natural language processing to capture and structure rationale from communications between software designers and prospective system users. Also, it should be mentioned that the approach proposed by Gruber and Russell (1996), i.e. retrospectively reconstructing rationale rather than attempting to record it, is also completely non-intrusive.

12.2.3 Rationale for Design Space Analysis and Deeper Reflection

We can distinguish two fundamentally different types of decisions that designers make: design-space decisions and other design decisions. The former decide what the features of the artifact will be; the latter do not. Much of the literature on design rationale in all fields has focused on design space decisions so exclusively as to give the impression that these are the only decisions designers deal with. But there are many other decisions that they make that do not directly decide system features but nevertheless have a profound, albeit indirect, affect on what features a system has. These decisions reflect a deeper level of reflection on factors that influence the design of the system.

The decision making processes associated with deeper reflection provide an important mechanism for improving the quality of software design. It is therefore important to understand the role that rationale approaches can play in supporting this sort of reflection. If a design process that software designers generally use does not explicitly support this sort of reflection, employing a rationale approach that promotes such reflection may enable the designers to improve the quality of their decisions.

12.2.3.1 Rationale for Design Space Analysis

The term *design space analysis* was coined by MacLean et al. (1996) to describe the sort of rationale-based decision making represented in their QOC (Questions, Options and Criteria) approach to design rationale. This approach documents the evaluation of proposed alternative answers to design questions. The questions they deal with are those whose answers represent features of the artifact being designed—usually software. Answering these questions amounts to making decisions about what features the artifact will have. To distinguish such questions from other types of questions dealt with in design, we will call the QOC-type questions *design-space decisions*. The *set of chosen answers* to all design space decisions in a project thus constitutes the complete *design* of the artifact. A crucial point for relating rationale to design is that design space decisions represent the points where rationale meets the representation of the artifact being designed.

Other decision-centric rationale approaches, such as IBIS (Issue-Based Information Systems) (Kunz and Rittel 1970) and PHI (Procedural Hierarchy of Issues) (McCall 1990), can also represent the rationale for Design Space Analysis, though their schema for representing the evaluation of alternatives differs from QOC's. Unlike QOC, IBIS and PHI can also deal with design questions that do not correspond to design-space decisions.

DRL (Decision Representation Language) (Lee 1990) resembles QOC in many ways, especially in its evaluation schema. Examples in the literature of the decisions that it deals with have been limited to design-space decisions; yet the author of the system (Lee) makes no claim about its use being restricted to design space analysis. It seems reasonable to give DRL the benefit of the doubt and assume that it can also be used for other kinds of design decisions.

12.2.3.2 Rationale for Deeper Reflection

Design-space decisions are not the only kind of decisions made in the part of the development process known as *design*. This can be seen by looking at decisions in the way that IBIS does: any decision to be made can be represented as a question to be answered; and any design question that needs to be answered represents a decision to be made. There are many important questions that can arise in design that do not have answers describing features of the artifact being designed.

Rittel listed a number of major categories of design questions, or *issues* as he called them (Kunz and Rittel 1970). These included the following:

- Factual issues – including questions about *what is, was or will be the case*
- Deontic issues – including questions about *what should be* or *ought to the case*
- Explanatory issues – including questions about *why something is the case* or *what causes something to be the case* or *what a term means* or *what effects something has*

The decisions on these issues typically do not directly describe artifact features, and yet they occur as part of the overall design effort and can decisively influence the design of the artifact.

An example of a factual issue would be, "Which rich Internet application (RIA) technology is likely to become dominant over the next five years?" An example of a deontic issue would be, "Should we be buying or building the graphics functionality that we need?" Examples of explanatory issues would include questions like "Why has security been so hard to achieve for previous versions of this software?", "How are we interpreting the meaning of the term *Rich Internet Application* in our project?" and "What does this requirement really mean?"

There are a number of different roles for non-design-space decisions in the larger design process. Often these roles are made clear by the circumstances in which such decisions arise. For example, some of these decisions arise from the attempt to generate decision alternatives. "How has this decision been made in other projects?" and "If cost were not a concern, what are all the conceivable ways we might try to accomplish this task?" are examples that have the role of helping to generate decision alternatives. Although both of these questions have answers that describe artifact features, they are not design-space decisions, because they *do not decide that the artifact should definitely have any specific feature.*

Often non-design-space decisions arise out of argumentative evaluation of decision alternatives. In collaborative design, for example, it is quite common for one designer to challenge a statement made by another; and sometimes these challenges get elevated to the status of non-design-space decisions that are decided by the whole group. This can happen because any statement that someone makes can be questioned, i.e. literally become the basis of a question about it. In fact, in any reasoning about a design space, questions can arise that are crucial for the design of the system but which *are not design-space decisions* in the sense of having answers that decide features of the artifact being designed.

Some non-design-space decisions take the form of meta-decisions, i.e. decisions about the decision making process itself. Examples of such questions include, "Which decision should we make first?" and "How much time should we allocate to making this decision?"

Decisions that occur during the design process but that are not design-space decisions gain there relevance to the design process only by influencing design-space decision making. The non-design-space decisions—such as the ones categorized by Rittel—are relevant to design only if they *inform design-space decisions*, i.e. influence the reasoning about design-space decisions. This indicates that, although rationale for non-design-space decisions is not *directly* about design-space decisions, it is always *indirectly* about such decisions.

A useful way of looking at the rationale for non-design-space decisions is that it represents *deeper reflection on the design process* than is represented in the rationale of design-space decisions. Rationale on the non-design-space decisions is especially important because it enables the deeper thinking about design that produces *more thoughtfully designed* artifacts that are of higher quality.

12.3 Specific Approaches that Integrate Rationale into Software Design

12.3.1 Rationale and Software Architecture

Software architecture is the one area of software engineering where rationale is most explicitly mentioned as an area of research. One reason may be the criticality of decisions made at this stage. As Bass et al. (2003) describe it:

> Software architecture manifests the earliest design decisions about a system, and these early bindings carry weight far out of proportion to their individual gravity with respect to the system's remaining development, its deployment, and its maintenance life. It is also the earliest point at which design decisions governing the system to be built can be analyzed.

The criticality of these early decisions indicates the grave importance that they be made with careful deliberation and remain consistent with the criteria identified by system stakeholders during requirements elicitation and analysis.

The importance of design decisions to architecture can be taken one step further—the architecture can be viewed as more than a collection of components and their relationships but rather as "a composition of architectural design decisions" (Bosch 2004). Tyree and Ackerman (2005) feel that architectural decisions are the key to "demystifying architecture products" and describe several places where traditional architectural approaches "break down." They point out that the lack of rationale results in system stakeholders continually needing to ask for answers to the same questions. These decision-centric views of software architecture lead naturally to decision-centric views on its rationale.

Not content with assuming that rationale is useful, Tang et al. (2006) surveyed architects to determine their opinions on the usefulness of rationale. The survey results showed that 85.1% of the architects surveyed considered rationale as important (4 or 5 on a 1-5 Likert scale). Other interesting results were that 74% of the architects did not remember the reasons behind their own design decisions and 80% agree that if the design rationale is not present they may not understand why a design was created without the assistance of the original designer.

12.3.1.1 Rationale and Architectural Decision Documentation

While it is encouraging to see rationale as a part of architectural design research, for the most part this information is delegated to a descriptive role where schema-based rationale appears in, and is defined by, a decision model or decision template. The information that populates the architectural knowledge repositories is for the most part provided manually by the designer as part of the architecture design process.

A number of architectural knowledge research projects stress the importance of rationale and capture it by including it as part of an architecture decision template that is filled in by the architect. Tyree and Ackerman (2005) proposed an architecture decision description template based on the Representation and Maintenance of Process Knowledge (REMAP) (Ramesh and Dhar 1992; Rhamesh and Dhar 1994) and Decision Representation Language (Lee 1990) rationale representations. The template expresses the decision, its status, assumptions, constraints, positions (alternatives) considered, arguments, implications, and related decisions, requirements, artifacts, and principles.

Templates are also used in the PAKME knowledge management tool (Ali Babar and Gorton 2007). PAKME stores design options, and their rationale, as "design option cases" that can be used to support case-based reasoning. A selected design option is represented as an architecture decision. Both design options and architecture decisions have rationale, also captured with the assistance of a template. Their rationale "describes the reasons for an architectural decision, justification for it, tradeoffs made, and argumentation leading to the design decision." (Ali Babar and Gorton 2007) and is described using a template (not described in the paper) based on those defined by Tyree and Ackerman (2005) and in the Views and Beyond (Clements et al. 2002) approach.

The Views and Beyond template is extended by Bass et al. (2006) by adding a causal graph of rationale that provides causal relationships between decisions. They also describe how a structural graph of rationale can capture rationale for each architectural element. The rationale focuses on the architectural elements' responsibilities with respect to achieving functional requirements and quality attributes. It also relates the architectural elements to the design decision alternatives. These two graphs provide two different ways of looking at the design—as a series of decisions (the causal graph) and as the result of making those decisions—the software structure.

An example of a model using rationale is the one developed under the GRIFFIN (a GRId For information about architectural knowledge) contract that structures software architecture project memories (de Boer et al. 2006). This model captures rationale as the alternatives proposed for a decision topic that are ranked based on concerns addressed in a particular viewpoint and that influence the decision topic. A major goal of their model is to "associate know-how, or rationale, with the know-what and know-how contained in design artifacts."

The Architecture Design Decision Support System (ADDSS) (Capilla et al. 2006) also utilizes a model that contains rationale. Their model (Capilla et al. 2007) contains several attributes that could be interpreted as rationale: the mandatory rationale attribute, recording the reason for making the decision, as well as several optional attributes providing alternatives, assumptions, pros/cons, quality attributes and a decision category. Another interesting optional attribute is iteration—this provides support for tracking decisions to the "architectural iteration" in which it was made.

Zhu and Gorton (2007) model design decisions using UML (Unified Modeling Language) (OMG 2005b) profiles and the Object Constraint Language (OCL) (OMG 2005a). Rationale is attached to the model by adding the rationale description as a UML tag (a mechanism for adding descriptive information to the model) on the relationship between design decisions and the non-functional requirements. Many believe that the

UML is the closest thing to a standard in software engineering and it is surprising that more approaches are not utilizing it for capturing rationale (although many approaches use it as a notation for describing their schemas).

Several approaches define their own ontologies, or notations, for rationale. The DAta Model for Software Architecture Knowledge (DAMSAK) (Ali Babar et al. 2006) defines design rationale as containing the following elements: description, comment, constraint, assumption, strength, weakness, cost, benefit, complexity, unresolved issues, justification, rule, context, tradeoffs, arguments, and other information, all text fields with the assumption of unresolved issues (which is an integer which must map to something else). Krutchen et al. (2006) define a rich ontology of design decisions. This ontology captures rationale directly as a textual description and indirectly through relationships between design decisions. The relationships defined in their ontology are quite comprehensive: constrains, forbids, enables, subsumes, conflicts-with, overrides, comprises, is an alternative to, is bound to, and is related to.

Archium (van der Ven et al. 2006) captures design decisions in a template-like format that either uses, or bears a strong resemblance to, Java annotations. The fields of their template are not called rationale, but captures alternative solutions along with their constraints, consequences, pros, and cons. There is also a place in the template to identify tradeoffs.

The Architecture Rationale and Elements Linkage (AREL) model (Tang et al. 2007) captures three types of rationale: qualitative rationale, arguments for and against design decisions; quantitative rationale that describes the costs, benefits, and risks of each design option, and a third type, the alternative architecture rationale to describe design options that were discarded. The qualitative rationale (QLR) is captured using a template and contains the issues, assumptions, constraints, strengths, weaknesses, tradeoffs, risks (and non-risks), the assessment and decision, and any other supporting information required to make a decision. The quantitative rationale (QNR) represents cost, benefit, and risk using an Architecture Cost Index (taking into account costs such as development, maintenance, and platform support), Architecture Benefit Index (which combines requirement priority and how well the decision satisfies it), and the Outcome Certainty Risk (how likely the architecture will be to meet its outcomes) and the Implementation Certainty Risk (the risk of implementation issues causing problems). The alternative architecture rationale (AAR) contains all the information in architectural rationale except that it is for alternatives that have been rejected (Tang and Han 2005).

12.3.1.2 Using Rationale to Support Software Architecture

The main emphasis of most of the approaches listed above appears to be using rationale descriptively as part of design decision documentation. There are approaches, however, that take a more prescriptive approach. For example, the PAKME architectural management system (Ali Babar and Gorton 2007) described above uses rationale as part of its "design option cases" that support case-based reasoning and the reuse of the rationale. This can be viewed as a means of *informing design*, by providing information that would be useful to the designers.

The Decision Goals and Alternatives (DGA) Design Decision Rationale (DDR) technique (Falessi et al. 2006) both supports design decision documentation and supports design decision-making. The DGA provides a decision documentation process that first has the decision-maker refine objectives, constraints, and sub-goals and a second stage that takes the designer through the "enaction" of decision phases where the designer assigns scores to the relevant attributes identified in the earlier phase. The scores provide the importance that an attribute has to the decision task at hand. This is an example of a rationale approach of *prescribing processes*.

The Architecture Rationale and Element Linkage (AREL) system described earlier can support several uses for rationale that support design modification and understanding, rather than assisting with the design activity itself. AREL has been extended to create the eAREL system (Tang et al. 2006) to support architecture evolution. This is done by storing a current version of each architecture element (AE) and architecture rationale (AR) as well as one or more "historical versions." The links between the between ARs and AEs provide traceablity forward, to perform impact assessment, backward, to provide root-cause analysis, and over time, to analyze the evolution of decisions and/or architectural elements. Change impact prediction to assess the effect of system requirements or decisions changing is also provided by a version of the system where AREL is modeled as a Bayesian Belief Network (BBN) (Pearl 1988) where the nodes of the network are architecture elements (requirements and decisions) or architecture rationale (the reason for making a decision) (Tang et al. 2005). In the AREL BBN, the links represent causal relationships. Architecture element nodes have two possible states: stable or volatile, indicating if it is likely to change, and architecture rationale nodes have the two states of valid or invalid. The conditional probability tables give the probabilities of different combinations of these states. The BBN supports two kinds of reasoning: *predictive reasoning* where the network is used to predict the effect of an architectural design change by changing the state of an architectural element to volatile, and *diagnostic reasoning* where a non-root node is set to volatile and the posterior probabilities of its ancestors are evaluated to determine possible causes for the change.

12.3.2 Strategies for Fitting Rationale into Architectural Design Processes

Almost all of the above-described approaches focus on fitting rationale into the processes of designing software architecture. But at first glance these approaches are so varied that it may seem hard to discover any common strategies for fitting rationale to design. A closer look, however, does reveal some basic trends. One of these is that more than half the approaches focus on the integration of rationale models with models of architectural artifacts to make hybrid rationale-artifact models.

There are two basic kinds of software architecture artifacts. One kind is the architecture itself and the various *design space decisions* that it consists of. As van der Ven et al. (2006) point out, these decisions are where rationale and architecture meet. This means they are also where rationale processes and the processes of software architecting must also meet.

The other kind of software architecture artifact consists of the many things that software architects create to do the work of design. These include things like patterns, tactics, scenarios, findings and design histories. A number of the papers integrate rationale and such artifacts into to hybrid models. PAKME does this for the stated purpose of integrating rationale into the processes of software architecting, and this appears to be the motivation for other approaches doing this as well. In general it seems that the more rationale can be tied to the creation of such artifacts, the more rationale processes can be fitted into the processes of designing software architecture.

Another trend is the reliance on the modeling of dependency relationships among decisions as a type of rationale. Sometimes this is little more than a re-invention of the dependencies found in rationale approaches like PHI, DRL and RATSpeak (Burge and Brown 2006). In other cases, a fundamentally different approach is taken to modeling dependencies. Most notably different is the AREL system's use of dependencies based on Bayesian Belief Networks. Regardless of how dependencies are modeled, the purposes for modeling them seem to be the same: traceability and predicting the consequences of change. These purposes are so important to SE that using rationale to model them fits rationale processes more closely to SE processes.

Other, more minor trends are the use of automatic capture of rationale and the introduction of more elaborate and quantitative modes of evaluation into rationale. The former seems aimed at reducing the potential conflict between rationale and software engineering processes. The latter bring evaluation of decision alternatives more in line with the types of evaluation used in SE.

12.4 Summary and Conclusions

Effective capture and use of rationale in software design requires that rationale approaches be skillfully fitted into the processes that software designers use. Arranging for a good fit is a complex undertaking that requires a detailed understanding of how approaches to representing, capturing and delivering rationale can support or conflict with software design. This chapter has used two methods for coming to such an understanding. The first was a theoretical analysis of various types and roles of design rationale and the way these affect software design. The second was a look at a variety of approaches that researchers have devised for fitting rationale into the design of software architecture. The theoretical analysis focused on the potential difficulties and benefits of rationale capture and use. The survey of research focused on the modifications of rationale schemas to include representation of SE artifacts, dependency networks and more elaborate modes of evaluation, all of which work to increase the fit between rationale and software design processes.

The complexity of the topic of fitting rationale into design and the great variety of approaches to doing so both suggest that much more research can and should be done on this topic if a broad consensus on approaches to design rationale is to be reached. At the same time, they make it clear that enormous progress has been made over the early days of rationale research when it was naïvely assumed that successful capture and use of design rationale in SE was a simple matter and that simple approaches would suffice. What is also clear is that researchers on design rationale have made progress not only in understanding the problems they face but also in solving them.

13 Rationale and Software VV&T

Designing and developing effective verification, validation, and testing strategies is always a challenge. The testing strategy needs to take into account the crucial balance between cost and quality and make appropriate tradeoffs depending on the specific project. In this chapter, we will investigate whether the presence of Software Engineering Rationale (SER) can assist in determining how and what to test.

13.1 Introduction

13.1.1 Verification, Validation, and Test

One of the most important parts of the software development process is Verification and Validation, or V&V. The goal of verification and validation, or V&V, is to provide an assessment of the ability of the software both meets its requirements and satisfies the needs of the user (IEEE 2004). Verification refers to assessing the software's conformance to its specification while validation refers to ensuring that the software fulfils the customers' expectations (Sommerville 2007). As Barry Boehm (1979; Sommerville 2007) puts it, validation asks "Are we building the right product?" and verification asks "Are we building the product right?"

The V&V process encompasses assessment, analysis, evaluation, review, inspection, and testing (IEEE 2004). Software, which includes documentation as well as code, can be assessed statically, through inspections or other analysis techniques, or dynamically through software testing. There are many reasons why inspection should be done in addition to testing. Inspection can find errors that might be masked by other problems during testing, can be performed before the software is complete, and can assess non-functional requirements like conformance to standards and the choice of algorithms (Sommerville 2007). Testing is still necessary to ensure the software runs and to assess performance, scalability, reliability, and other qualities during operation. Testing continues to get more and more difficult as increasingly powerful systems (at the same price) provide the ability to run increasingly complex software (Stobie 2005).

The process of ensuring that the software conforms to its specification, as well as the evaluation of the development process itself, falls under the category of software quality assurance (SQA) (IEEE 1990). Quality assurance involves planning for how quality will be achieved and measured. QA strategies vary between organizations with some planning for quality from the start and monitoring progress while others viewing SQA as simply being testing, an approach to SQA compared to "locking the barn door after the horse has escaped" (Baker 2001).

13.1.2 Software Testing Issues

One of the most famous software engineering phrases comes from Dijkstra: "program testing can be a very effective way to show the presence of bugs, but it is hopelessly inadequate for showing their absence" (Dijkstra 1972). This highlights the primary difficulty of software testing – it is impossible to insure that there are no errors. The presence of errors is inevitable. Error probabilities for experienced programmers are around 1% (one in 100 lines of code) and increase for less experienced developers (Wang and Tan 2005). Testing can comprise a significant percentage of the development effort (Juristo et al. 2006).

There are many decisions that need to be made when developing a testing strategy. One is the level of software integrity required. IEEE 1012-2004 (IEEE 2004) defines four integrity levels ranging from level four, where if the software does not run correctly there may be significant consequences (loss of life, equipment, or money) to level one where the consequences of failure are minor. Since software testing can be very expensive and time consuming, it is important to determine how rigorous the effort should be and if there may be more important considerations such as time-to-market that should take precedence.

Decisions also have to be made about what types of testing to perform. Some researchers break testing down into two types: debug testing and operational testing (Frankl et al. 1997). This fits with two primary reliability testing goals – to find and remove defects in the software and to evaluate the ability of the software to operate as expected. Debug testing is more effective at finding defects but may result in focusing testing resources in finding problems that may never appear operationally. Operational testing evaluates the system under realistic conditions but may not find rare, but possibly catastrophic problems that may occur under extreme conditions.

The number of tests, or test cases, is also an issue where cost/reliability tradeoffs must occur. Exhaustive testing of all possible inputs is almost always prohibitively expensive. Test cases need to be chosen carefully so the level of testing is appropriate for the integrity level for the system as well as achieving sufficient test coverage.

13.1.3 Objectives of this Chapter

This chapter provides a brief introduction into some components of software VV&T and describes some ways that rationale could contribute to each of them. That description is then followed by a discussion of how rationale would support software testability, test case prioritization, and component selection and test. We then conclude by examining how rationale captured for test planning and strategy selection could be used to support future development efforts.

13.2 Types of Software VV&T

13.2.1 Inspection

Inspection, of code or other software artefacts, is an important component of the VV&T process. As stated earlier, inspection can find problems that testing often can not. Inspection can also be employed at the early stages of system development to detect specification and design errors. If these errors can not be detected and removed, the resulting system is likely to be poorly structured with faults due to the design flaws (Kitchenham and Linkman 1998).

Numerous studies have been performed to compare the effectiveness of inspections versus testing. For formal inspections (also known as Fagan Inspections), the inspections have been shown to be 7.4 times more productive (looking at the ratio of errors found to effort expended) than testing (Eickelmann et al. 2002).

Inspection and rationale can work together in a number of ways. As with exhaustive testing, exhaustive inspections may not be feasible for many projects. If rationale is available, it can be used to help determine where the inspection efforts should be focused. The rationale for a software system can point out where the non-functional requirements best addressed by inspection (such as maintainability) were used to drive the decision-making process. The rationale can also point out where these non-functional requirements were not considered. If maintainability, for example, was not considered for some software artefacts and it should have been, those artefacts should be inspected.

Inspections can also be used to assist in rationale capture by requiring that the rationale be inspected along with the software artefacts it applies to. The inspectors can ask questions such as:

- Where there alternatives considered?
- Were criteria used to drive the decision-making process appropriate? Were items given the correct priorities relative to overall system goals?
- Were there any assumptions made that need to be documented in the rationale?

Incorporating rationale capture into the inspection process makes rationale capture an integral part of the development process. It also helps to focus the collection on areas that were considered crucial enough to merit inspection. Also, since inspection is a collaborative activity, it ensures that the decisions made are reviewed and ensures that input from team members other than the primary developer are taken into consideration. Several rationale based systems have been shown to be helpful in keeping meetings on-track and in supporting collaboration and negotiation. Examples include WinWin (Boehm et al. 2006) for requirements negotiation and Compendium for meeting facilitation (Buckingham Shum et al. 2006).

13.2.2 Unit Testing

Unit tests are tests performed on the "smallest possible testable software component" where units can be classes, small commercial-off-the-shelf (COTS) components, in-house components, or procedures/functions (Burnstein 2003). These tests are typically performed by the software developer. Unit testing consists of three phases: planning, test set acquisition, and test set measurement (execution and evaluation) (ANSI/IEEE 1987). In Test Driven Development (TDD) (Beck 2002), also known as Test-First Development, the unit tests are written before the code. The ability to do this is supported by unit testing frameworks like JUnit (http://www.junit.org).

For unit testing, one key decision is how many tests to write. As with other forms of testing, exhaustive testing is not possible. The techniques of white (or glass) box testing can be utilized – tests should be written to cover all paths, as well as black-box testing – creating equivalence classes to test different types of inputs and to ensure that boundary conditions are covered. Test case prioritization is also important – running higher priority unit test cases first has been shown to increase fault detection rates (Do et al. 2006) (Rothermel and Elbaum 2003).

13.2.3 Integration Testing

Integration testing refers to the testing that is performed as the various components that make up the system are combined. There are many different methods that can be used during integration. The method chosen depends on the process used to develop the system and when components are available for integration. As a general rule, the "big bang" approach where all components are integrated together at once is something to be avoided because that approach makes it more difficult to isolate what component is responsible when an error is detected (Schultz 1979).

Alternatives to the "big bang" include bottom-up testing where lower level components are tested first, top-down testing where the higher level components, starting with the user interface, are tested first, and other variants, such as sandwich testing, which performs testing of the top and bottom levels in parallel using the components in the middle layer (Bruegge and Dutoit 2004). One heuristic used to evaluate different integration strategies is to ensure that the "most important components" receive the most testing.

The rationale can help with determining the integration strategy in several ways. One is by its support for requirements traceability. If requirements are used as arguments for selecting decision alternatives that then map to the code implementing these alternatives, this mapping can show what portions of the software apply to which requirements. This is useful to know when assessing the criticality of components. Non-functional requirements can also appear in the rationale and can be used to assess criticality.

13.2.4 System Testing

System testing is a general category that refers to a variety of different types of tests that could be performed on the system as a whole. One type of system testing is the Acceptance Test – tests that demonstrate to the customer that the system functions as required.

Regression testing is another form of testing that works with the system as a whole. Regression tests are performed on a system when changes are made during software maintenance. The goal is to ensure that new changes added do not break any of the existing functionality. Regression testing can be very expensive so it is often necessary to only perform a subset of the possible tests. The use of rationale to support regression testing is described later in this chapter.

There are also a number of specialized tests that could be performed on the completed system. These could look at aspects such as system performance, reliability, security, and other non-functional system characteristics. Performance alone contains many different sub-tests including stress testing (testing the number of "requests"), volume testing (data amount), security tests, timing tests (checking timing constraints), and recovery tests (the ability to recover from failures) (Bruegge and Dutoit 2004). The rationale can be used to help determine how the different types testing should be performed by identifying areas of the system where those aspects were considered during development. This assists in targeting the tests to areas where failures are likely to be the most critical.

13.3 Rationale Support for Software VV&T

13.3.1 Rationale and Testability

Since testing is both difficult and costly, developers should be thinking about how testability could be designed into their software The IEEE Standard Glossary of Software Engineering Terminology (IEE 1990) defines testability as follows: "The degree to which a system or component facilitates the establishment of test criteria and the performance of tests to determine whether those criteria have been met." Knowing the testability of a software component can help guide testing by indicating how difficulty it will be to find defects in that component and using that information to determine the "testing intensity" (Voas and Miller 1995).

Rationale can support testability in two ways. The first is by encouraging the developers to provide rationale for their decisions and by offering testability attributes as reasons for considering one alternative over another. Boehm et al. (1979) included testability characteristics in their Software Quality Characteristics Tree. The characteristics identified are communicativeness (understandable inputs and outputs), self-descriptiveness (well documented, traceable), and structuredness (well organized). Bass et al. (2003) define testability "tactics" that can be used to design for testability at the architecture level. These include information recording and playback capability, separating the interface from the implementation, providing specialized interface for use by test harnesses and building in monitors to save state information. Burge (2005) included testability attributes in the SEURAT Argument Ontology such as function visibility, minimizing variable re-use, providing triggers, supporting instrumentation, and providing re-entry points. If the developers use a rationale support system to assist in their decision-making this could increase awareness of testability criteria as reasons for their decisions. In addition, the presence of the rationale would reduce the risk of software maintainers making design decisions that limit testability by modifying the software in ways that conflict with prior testability goals. Since functionality supporting testability, such as instrumentation and re-entry points, tends to cross-cut much of the system functionality the ability to trace from the code to the traceability goal is especially useful.

The availability of rationale describing the testability of the software can then be used to help set the testing intensity as described by Voas and Miller (1995). If the component has been designed to be easier to test it may require fewer test cases to provide sufficient confidence in its correctness. If the component did not consider testability and the rationale indicates that it is critical, this would indicate that a higher testing intensity would be warranted.

13.3.2 Rationale and Test Case Prioritization

A key component in both incremental development and in software maintenance is regression testing – repeating earlier tests to ensure that new modifications have not harmed existing functionality or quality. Regression testing can be both time consuming and expensive so it is often not possible to repeat the entire set of tests for each modification. There may even be cases where if the new modification is critical enough, such as an emergency patch for a fatal error or security flaw, where time is limited and the number of tests run is significantly constrained (Srivastava and Thiagarajan 2002). The need to perform regression testing both effectively and efficiently requires some form of test case prioritization. A second goal of prioritization is to find problems as early as possible during regression testing so they can be corrected.

There are a number of testing goals that can drive test case prioritization. Some examples include achieving test coverage quickly, testing frequently used features first, and early defect detection (Rothermel and Elbaum 2003). Studies have shown that prioritization techniques consistently outperform randomized testing (Rothermel and Elbaum 2003).

The presence of rationale can supplement methods for test case selection by providing a link between the software and the quality attributes that need to be assessed when changes are made. When a change is introduced into the code base, the quality attributes used to drive the original decisions can indicate the qualities that need to be evaluated after a change has been made. Traceability links between functional requirements and the code, also captured in rationale, can help in the determination of which functions are most critical and should receive re-testing the earliest in the test suite.

13.3.3 Rationale, Testing, and Component Selection

Many software development projects are making use of re-used or purchased components. While there are many advantages to this approach, it does come with some additional difficulty in software testing. One goal of COTS or component-based development is to reduce costs and increase reliability. This is dependent on the reliability of the components chosen. While this software may have been extensively tested, that testing may not have been performed under the same circumstances as it will be used in a new product. For a component to be trusted, the component-provider needs to have been tested in all possible configurations and independent of a specific context of use (Harrold 2000).

Rationale can be of considerable support during the component selection and evaluation process. Rationale is often described as the way to capture intent (Sim and Duffy 1994). The success of COTS and component-based development efforts hinges on the compatibility of the intended use of the component from the component-provider and the component-consumer perspectives. If the rationale for component development is provided, that information would be invaluable to the consumer. Avoiding selection of a poorly suited component is the first step in avoiding component introduced defects.

Assuming that the component appears to be suitable, the rationale can also be of assistance if the rationale for the component-provider's testing strategy is available. Did the tests focus on specific qualities such as performance and reliability? Has the component provider designed their component to support ease of testing? Are test harnesses for the component provided? The more information available about how the component was tested, the higher the consumer confidence will be. The component consumer can examine the rationale behind the testing strategy used by the provider to ensure that it is a good fit with what they need and expect from the component.

13.4 Software Testing Rationale

Quality Assurance and software VV&T require careful planning in order to determine appropriate strategies and tools. As with the rest of software development, the decision problems encountered, the alternatives considered, and final decisions made can be captured as rationale. Bertolino (2007) identifies six questions identifying any approach to software testing: why (the test objective), how (test selection), how much (test adequacy), what (levels of testing), where (testing context), and when in the product lifecycle. The answers to these questions form the testing strategy and can be captured in the testing rationale.

13.4.1 Testing Rationale

A good VV&T strategy is unlikely to be something that just happens – there needs to be careful planning to determine what the testing priorities are and how to assess test effectiveness. As with other types of planning, there will be many decisions made that can be captured in the rationale. Capturing test development rationale is useful in negotiating priorities for the current development effort and also in determining testing strategies for subsequent efforts. The combination of test history and rationale for the strategies chosen would be very helpful in determining if those strategies were successful and how they should be modified.

As mentioned earlier, there are many tradeoffs that must be made during VV&T planning. For example, what level of testing integrity (IEEE 2004) is required? Is time to market more important than quality? It is important that projects determine what quality means for them and what costs they are able to incur to achieve it. If pursued incorrectly, quality can "destroy value" of a product (Favaro 1996).

Test case selection is another area where decisions need to be made. What is an appropriate granularity level for test cases? Small cases are easier to prioritize but add to the cost of test suite management (Rothermel and Elbaum 2003). The rationale is a natural place to capture these and other testing tradeoffs. Similarly, the choice of integration strategy and which types of system tests are run are also decisions that should be justified in the rationale. Rationale for inspection decisions can also be captured—what was inspected and using which technique? There are many different techniques for reading code all of which are based on their own assumptions about code inspections (Thelin et al. 2003). The rationale can record the reasons for the choice of technique so those reasons can be compared to those of the project's overall QA strategy.

Rationale should also capture the reasons behind the choice of metrics collected. While metrics are a valuable mechanism for assessing software quality, they can be expensive and time consuming to collect. The value of metrics in achieving quality goals and monitoring high risk areas needs to be balanced against the cost of collecting the metrics data (Clapp 1993). The project goals and their relationship to the metrics collected can be captured in the rationale. The type of metrics collected may also be influenced by the tool support available to aid in collection. This information should also be collected in the rationale so that if tool availability or preferences change it will be easy to determine if this should result in a corresponding change in which metrics are collected.

13.4.2 Uses for Testing Rationale

Rationale capture can be viewed as a form of knowledge management where one of the primary goals is to capture the expertise utilized during a software project. While valuable during a single project, a key goal is in being able to share knowledge between projects. One way towards this goal from the testing perspective would be to perform a retrospective examination of testing strategies with the goal of applying the lessons learned to future projects. For example, test case selection is likely to involve making assumptions about how effective each case will be at finding defects (Chernak 2001). Analyzing the actual effectiveness and comparing that against the initial assumptions will help determine if those assumptions were valid. That information can then be fed into test planning on subsequent projects.

Testing rationale can also be a valuable contribution to the QA process for a project. The rationale can be evaluated to determine if the reasons behind the testing decisions were consistent with overall project goals. It can also be used to evaluate how well supported the selected decision alternatives were. Requiring that rationale be collected for testing decisions to be used as part of the quality assessment may encourage test planners to put more thought into choosing their strategy than they might have otherwise.

13.5 Summary and Conclusions

Software VV&T is crucial in ensuring the quality of delivered software. Developing and carrying out an appropriate strategy can be both difficult and expensive. There are many decisions that need to be made regarding what needs to be testing and how that testing should be performed. Not all tests are of equal value and the costs and risks need to be examined as part of a "value-based" testing strategy (Ramler et al. 2006).

Rationale can support software testing as both an input into and an output of VV&T. As an input, the rationale provides the intent behind the software decisions. This could highlight potential weakness that should be evaluated as well as indicating some of the types of tests that might be appropriate given the non-functional requirements that guided specific decisions. The testability of a piece of software can be explicitly captured in the rationale. As an output, rationale for decisions made on how and what to test can be later compared with the eventual outcomes of that testing to provide insight into how testing processes can be improved on future projects.

14 Rationale and Software Maintenance

Software maintenance can be very expensive part of the software development process. Anyone working in the software industry during the years leading up to the year 2000 (Y2K) is all too familiar with the often unexpectedly long life-span of many software systems. The difficulties of maintaining these systems are acerbated because the original developers are often not available. Software Engineering Rationale (SER) would provide insight into why the system is the way it is by giving the reasons behind the decisions made during design and implementation. Rationale could help to indicate where changes might be needed during maintenance if design goals change and help the maintainer avoid repeating earlier mistakes by explicitly documenting alternatives that were tried earlier that did not work. In this chapter we will look at these and other ways that rationale can assist with software maintenance.

14.1 Introduction

14.1.1 Software Maintenance and Evolution

Software maintenance refers to "the modification of a software product after delivery to correct faults, to improve performance or other attributes, or to adapt the product to a modified environment" (IEEE 1998). This process is often referred to as software evolution, although evolution can be considered to be only one phase in a software maintenance cycle that also includes servicing (minor changes made when the system is no longer capable of being evolved), phase-out, and close-down (Bennett and Rajlich 2000). Lehman's laws state that if a system is not evolved, it becomes less satisfactory to the users and is perceived to have declining quality (Lehman 1996).

If a software system is successful, it could potentially spend a large percentage of its lifetime in the maintenance stage and maintenance costs could be significant. Costs appear to be increasing over time with life-cycle costs devoted to maintenance rising from 40% in the early 1970s to 90% in the early 1990s with 80% of these costs going for system improvements (Pigoski 1997). Despite its importance, this is an area that requires more attention. Unfortunately, software development is often driven by cost and schedule with no incentive to the software developers to build maintainable software (Pigoski 1997). In 2001, a study was performed to study process improvement efforts in industry (Hall et al. 2001). Qualitative data was collected from thirteen companies using focus groups and quantitative data was collected from 85 companies using questionnaires. The study showed that formal process improvement models did not sufficiently address maintenance. Kajko-Mattson (2001) and a class of software maintenance students studied eighteen organizations in Sweden to assess how well they met a set of documentation requirements, many of which were intended to support software maintenance, and demonstrated that there needed to be improvement. Problems with software documentation, which is often out of date, may be why software maintainers often do not trust the documentation and get most of their information from the source code (Singer 1998).

One way to potentially decrease the cost and risk of software maintenance would be to capture and use the rationale behind decisions made during design and implementation. Maintainers would no longer need to guess at the developers' intent but instead could take advantage of developer knowledge when making maintenance decisions.

14.1.2 Objectives of this Chapter

This chapter describes the software maintenance process and how it is supported by rationale. It defines the types of software maintenance and then focuses on two main areas: how maintenance can be improved and on how maintenance can be supported. Maintenance improvement involves both designing the code to be more maintainable and re-engineering existing code for maintainability. Maintenance support involves predicting where maintenance will be required, evaluating the impact of proposed maintenance changes, understanding the software being maintained, and studying the history of how the software has evolved over time.

14.2. Types of Software Maintenance

Software maintenance involves maintaining more than just the code. The most commonly mentioned types are the four given by Lientz and Swanson

(1980): corrective (repairing faults), adaptive (changes that do not add functionality but adapt the software to changes in the environment), perfective (updates to add functionality), and preventative maintenance (changes to make the software more maintainable in the future). Chapin (2000) took a wider approach and identified twelve types of software maintenance: training, consultive, evaluative, reformative, updative, groomative, preventive, performance, adaptive, reductive, corrective, and enhancive. The first five types do not involve modifying the software but instead affect how the stakeholders or developers interact with it (training, consultive, evaluative) or update the software documentation (reformative, updative).

For any type of maintenance, there are several activities that need to take place. Kitchenham, et al. (1999) lists four activities: investigation, modification, management, and quality assurance. Investigation involves impact assessment to determine what impact the change will have, modification is the change itself, management encompasses all management activities including configuration management, and quality assurance includes testing and other activities that must take place to ensure that the changes do not damage product quality. Figure 14.1 shows questions arising during these activities that could be answered by the rationale.

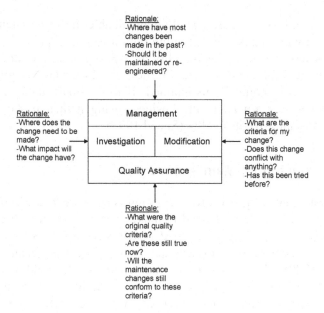

Fig. 14.1. Rationale Support for Maintenance Activities

Software maintenance changes can be needed for many reasons. Corrective maintenance is necessary when a problem is detected in the software. Enhancive maintenance is necessary when requirements are added to meet customer needs. Another reason for making changes involves assumptions. Over time, assumptions that were made during development can become invalid. This invalidation is a major driver for software evolution (Lehman 2005). When these changes are made, it is important to understand how they impact the software and its ability to meet customer requirements. Rationale can assist with this by relating the code being modified to the requirements and assumptions that drove its design.

The software maintenance process is also affected by who performs the maintenance. In some cases, this is not the original developers. It is not uncommon on large projects to award separate contracts for operations and maintenance (OEM). If the maintainers do not have access to the original developers it is quite possible that the rationale will provide the *only* insight into the original developers' intent.

14.3 Improving Maintainability

Not all software systems are equally maintainable. Ideally, systems should be designed with maintenance in mind with a well documented, easily extensible design. Using maintainability as a design goal and documenting it in the system rationale is a step in the right direction. Not all systems, however, can or should be maintained. If the system quality is low but the system is still crucial to the business using it then it should be re-engineered to improve its quality (Sneed 1995).

14.3.1 Designing for Maintenance

Many of the goals of good software design, such as reducing coupling and increasing cohesion, are intended to make it easier to extend the software more easily. Studies have shown that system structure does have an effect on the time required for and accuracy of software maintenance (Gibson and Senn 1989). This indicates that maintenance costs could be reduced if the software is designed so that it can be more easily modified later.

The ability to easily extend software was one of the driving forces behind Design Patterns (Gamma et al. 1995). Design Patterns are solutions to common problems that reduce or isolate dependencies between classes. Examples of patterns that are especially valuable during maintenance are the Facade pattern, which isolates clients from the code that provides services to them by providing a single class as an interface, and the Adaptor pattern, which creates a "wrapper" around the interface of a component or system so that clients only access the controlled wrapper interface, and the Mediator pattern, which uses a single Mediator class to control how underlying classes work together so that they are not dependent on each other's interfaces.

One of the goals of writing extensible software is to follow the "Open-Closed Principle" (Myer 1988). The open-closed principle states that software should be "open for extension" but "closed for modification", i.e. it should be possible to extend the software without modifying existing code. A study performed using the State design pattern showed that following it *correctly* results in code that follows the Open-Closed principle (Ng et al. 2006). Design patterns can result in more complicated designs. Prechelt et al. 2001) performed an experiment to determine if maintenance time was reduced by using a pattern rather than a simpler solution. In most tasks studied, the pattern was shown to be beneficial but there were cases where the simpler solution had fewer errors or took less time to maintain.

Rationale can contribute toward better designed software in several ways. One is in the selection of design patterns. The Design Recommendation and Intent Model Extended to Reusability (DRIMER) system used rationale to assist with design pattern selection and adaptation (Peña-Mora and Vadhavkar 1997). Rationale is also used to drive design when applying the NFR Framework (Chung et al. 2000). In the NFR Framework, the Adaptability NFR (non-functional requirement) was used as the driving force to design adaptable software architectures (Subramanian and Chung 2001). This process involves considering multiple design alternatives and uses and records the rationale. The NFR Framework was used in the SA[3] (Software Architecture Adaptability Assistant) tool to develop adaptable architectures (Subramanian and Chung 2002). The design tradeoffs and rationale are a critical component in evaluating which alternative architecture is most suitable.

Rationale also plays an important role by documenting where maintainability (and the related NFRs of flexibility and adaptability) was involved in decision-making. This information will explain the design and implementation to the maintainer and help to prevent changes that conflict with those goals. An example would be documenting where, how, and why a design pattern is used so that the pattern is not inadvertently broken by later development.

14.3.2 System Reengineering

Lehman's second law states that systems being evolved become increasingly complex unless something is done to reduce that complexity (Lehman 1996). Fowler et al. describe "bad smells" as potential problems in code, or code structures, that are candidates for refactoring (Fowler and Beck 1999). Refactoring involves removing the bad smells to improve the design of the code. Examples of bad smells include duplicated code, methods that are too long, and complex conditionals. Another response to the problem of code deterioration is to built preventative measures into the development cycle. The Class Deterioration Detection and Resurrection (CDDR) activity and Code/Class Growth Control (CGC) activity can be applied at each process iteration to address problems of high coupling and duplicated code (Subramaniam 2000).

System reengineering involves re-writing legacy systems to either increase their maintainability, port them to a different platform, increase reliability, prepare for modifications, or any or all of the above (Sneed 1995). The reengineering process includes several tasks: reverse engineering the existing system (re-capturing models), determining what repairs need to be made to the structure, and updating the legacy system (Nierstrasz et al. 2005). Demeyer et al. (2003) described these tasks in detail in a series of "re-engineering patterns" that describe approaches to reverse-engineering the code, testing to support evolution, and migrating from the legacy systems to re-engineered systems.

Reengineering usually does not modify the function of original system or change its architecture (Sommerville 2007). In some cases, however, more drastic changes are desirable. One example is when reengineering is performed to migrate the legacy system from its current architecture towards a component-based one. Mehta and Heineman (2002) developed an approach to make the software more maintainable by transforming it into fine grained components where features that change frequently can be isolated. The components corresponded to system features which were identified by examining the system's regression test suite. Code profiling can be used to detect which code is executed when testing which features.

Rationale can support system reengineering in several ways. One is in the negotiation that should take place to determine the advantages and disadvantages of reengineering. In some cases, it may be more efficient and economical to build a new system rather than reengineer an existing one. These arguments and the tradeoffs required can be captured in rationale.

Once the decision has been made to reengineer, if rationale is available for the legacy system it would assist with code comprehension by documenting what decisions were made and why. The rationale may also capture the original intent behind the decisions and can indicate where changes may need to be made if the system has been changed to no longer meet that intent.

14.4 Software Maintenance Support

14.4.1 Maintenance Prediction

Maintenance prediction involves determining what portions of the software system are likely to require changes in the future. This information is valuable because it indicates where developers should concentrate effort towards making the system more extensible. This information is also helpful in planning maintenance releases.

Stark et al. (1999) studied 44 software releases for seven products to study the type, frequency, and impact of requirements changes during software maintenance. In this study, any approved change request was considered to be a requirement. They generated a taxonomy of requirement change types and collected data on the source of the change, when in the development cycle it was requested, and the time required to make it. They discovered that it was useful to obtain the *intention* of the requirement (the rationale behind it). The requirements taxonomy and the historical information about how long they took to implement them can be used to estimate the time needed for future changes of the same type and to assess the amount of schedule slip occurring if changes were made to scheduled releases. Information about when in the cycle requirements changes took place was also helpful in controlling change.

14.4.2 Impact Assessment

When performing software maintenance, it is important to understand the impact of any changes proposed. The change needs to be assessed to determine the size, and therefore the cost, of the change. The change also needs to be assessed to insure that it is consistent with system requirements, both functional and non-functional. It is not uncommon for proposed changes to conflict with each other or with the original system goals.

The rationale for the system can assist with impact assessment. The Software Maintenance Expert System (SMES) (Avellis et al. 1993) is a blackboard architecture based system that uses an Assumption-based Truth

Maintenance System (ATMS) (deKleer 1986) to evaluate the impact of changing a design decision on the rest of the system. In SMES, each design decision is linked to a design plan that implements it.

More recently, the SEURAT system (Burge and Brown 2006) was developed to perform impact assessment of changing requirements and assumptions using the system rationale. SEURAT evaluates support for design alternatives based on the requirements, assumptions, and non-functional requirements that argue for and against them. If a requirement or assumption is disabled, the support for the alternatives is re-evaluated and the maintainer is alerted which decisions may require changes. Non-functional requirements, stored in an argument ontology (Burge 2005), each have a priority associated with them which can be modified on a global level or for a specific decision. This modification will trigger a re-evaluation and the maintainer alerted if a change might be needed.

Impact analysis, and how it can be supported by rationale, is described in more detail in the Change Analysis chapter of this book.

14.4.3 Program Comprehension

In order to successfully maintain a piece of software, the maintainers need to be able to understand it. Program comprehension can be a difficult process, especially for large software systems. When fixing a bug, understanding new code can take between 70% (for experienced developers) and 90% (for new programmer's) of the programmer's time (Eick 1998). When given a maintenance task, the maintainer needs to find the code relevant to the task, learn their dependencies, and add or update the needed code (Ko et al. 2005). Maintainers typically use "beacons"—useful code fragments, comments, or variable/procedure names—to find their way through the code. Novice programmers, however, do not recognize beacons (Crosby et al. 2002) and that adds to the difficulty of the maintenance task.

Some program comprehension approaches attempt to help with comprehension of the system as a whole. The LaSSIE (Large Software System Information Environment) environment (Devanbu et al. 1991) provides access to the software via a number of different viewpoints by making use of intelligent indexing and a domain model. The construction of the knowledge base required by LaSSIE, however, is a manual process. Evolution is also supported by developing software that incorporates Intentional Views (Mens et al. 2002). Intentional views group software into "concerns" using naming conventions and inheritance. Prolog rules are used to capture those conventions and use them to extract the software for each view. Automated clustering of software components is supported by the PROCSSI system (Maletic and Marcus 2001) which uses Latent Semantic Indexing (LSI) to compute similarity using variable names, type names, and comments. Concept analysis (Siff and Reps 1997) is used to infer repeating design patterns from code (Tonella and Antoniol 1999). This approach finds groups of classes with similar structural relationships to each other and can be applied to find patterns in code without having a pre-defined pattern library.

Software visualization techniques can be used to support maintenance. Software visualization can be defined as "the use of the crafts of typography, graphic design, animation, and cinematography with modern human-computer interaction and computer graphics technology to facilitate both the human understanding and effective use of computer software" (Price et al. 1998). Software visualization environments assist with visualizing the structure of the program, via call-graphs and other views, and the behaviour. Behaviour visualization can be useful in looking for system bottlenecks as was demonstrated by the PV (Program Visualization) prototype which would display execution time and memory use by system components (Kimelman et al. 1998). Statistics computed from the change history of the code can be useful in maintaining very large computer systems. The SeeSoft visualization technique (Eick 1998) uses colors and graphics to visualize changes made to source code. This can aid in detecting duplicate code, and determining which code was modified the most frequently and most recently. Animations can be used to view changes over time.

Other approaches are designed to help maintainers find the relevant code. This is related to the concept assignment problem (Biggerstaff et al. 1993) where program code structures need to be mapped to the human-oriented domain concepts that they implement. This is a significant difficulty during maintenance. A study of corrective and perfective maintenance showed that on average programmers spent 25 ($\pm$ 9) minutes out of 70 total inspecting code that was not relevant to their task (Ko et al. 2005). Robillard (2005) developed an algorithm, implemented in an Eclipse PlugIn, that takes an initial set of task-related elements and returns other program elements related to that set ranked by interest to the developer. Feature location approaches can also assist with this process by mapping features (user visible sets of requirements) to the source code. This can be approached using static techniques which use dependency analyses but do not execute the code and dynamic techniques which examine what code was executed when running test cases. Static techniques tend to be imprecise while dynamic techniques only capture code relevant to the specific inputs given to the program (Koschke and Quante 2005). Hybrid techniques combine static and dynamic approaches. Other methods for finding relative code include program slicing (Tip 1995), information retrieval techniques (Antoniol et al. 2000), and data mining over software repositories (Zimmermann et al. 2004).

Rationale can assist with program comprehension in several ways. Rationale can serve as a way to index from requirements to the code and from the code back to the requirements. Forward tracing, from requirements to code, can be used to examine where the different requirements are implemented. Reverse tracing, from code to requirements, illuminates the intent of the implementation. The relationship between the code and its requirements is crucial in order to ensure that changes made during maintenance do not introduce requirements violations.

When performing specific maintenance tasks, the maintainer needs to determine which code is relevant to the problem being solved. Using the rationale to index into the relevant source code can significantly decrease the time required, especially for non-expert developers (Burge and Brown 2006). In a study performed using SEURAT (Burge 2005), novices Java programmers without rationale assistance spent significant amounts of time trying to find the code they needed to modify and were often side-tracked by class and method names that appeared to be applicable but were not.

When von Mayrhauser and Vans (1994) studied program comprehension processes they listed one of their information needs as the ability to obtain connected domain information and suggested that this be provided using the design rationale and its ability to connect the application's algorithms to the application's purpose.

14.4.4 Maintenance Recovery

The rationale behind changes made to the software over time is required for what we will call maintenance recovery—the ability to document and track maintenance changes. While most reengineering efforts look at the latest version of the software, the history of how the system has evolved over time indicates where "chronic problems" are located (Nierstrasz 2005). In addition, understanding how a system has evolved can help predict how it may evolve in the future (Antón and Potts 2001). Software metrics can be used to analyze the software to detect evolution-critical parts (code that is likely to require evolution because of poor quality), evolution-prone parts (typically because they correspond to volatile software requirements), and evolution-sensitive parts (software that is likely to break during evolution, typically due to tight coupling) (Mens and Demeyer 2001). A combination of metrics and software visualization techniques are used in the Evolution Matrix (Lanza 2001). The Evolution Matrix can be used to visualize changes in the size of systems and classes throughout multiple evolutions. The developers of the Beagle maintenance support tool (Godfrey and Tu 2002) are particularly concerned with understanding invasive change—changes made to a software system that involve significant changes to the systems structure. Beagle supports origin analysis which determines if a software entity that is in a new version of the software but not the old one is actually new or if it is an entity (renamed, moved, or modified) from the earlier version. This information can then be used to build the "evolutionary history" of the application.

One source of information describing changes made during evolution is the change history extracted from configuration management tools. This information can be used to both capture the maintenance history and, if the reasons behind the changes are documented, the rationale. When evaluating a systems history it is important to not only know what changed, but why. The differentiation between evolution-critical, evolution-prone, and evolution-sensitive portions of the code will be easier if the reasons why these parts of the system could be extracted from the rationale as well as being inferred from metrics.

14.4.5 Maintenance Rationale

Rationale can, and should, be captured during maintenance. One reason for doing so is so that the rationale for maintenance changes can be compared for rationale for the decisions made during earlier changes to check the consistency of decision-making criteria. If the criteria differ, that could signify that new changes may have an adverse affect on the system quality as identified by the original developers or it could indicate a priority shift that may necessitate revisiting other earlier decisions.

Rationale can also support collaboration in maintaining very large systems by allowing maintenance knowledge to be shared between team members. Loughher and Rodden (1993) built a documentation system to support capture and sharing of maintenance rationale. They did not use an argumentation approach to their rationale because they believed that maintenance rationale is more focused on explanation than deliberation. Their system worked by allowing source code annotation using a markup-language to link "maintenance comments" to the code. Maintenance comments can be in text form or simple graphics, such as a flow charts.

The Cooperative Maintenance Conceptual Model (CM^2) (Canafora et al. 2000) also captures maintenance rationale. The goal of CM^2, and Cooperative Maintenance Network Centered Hypertextual Environment (COMANCHE), the system that uses it, is to support collaboration over an extended period of time by making rationale available to future maintainers. In the CM^2 process, maintenance starts with a maintenance request. The rationale for the design for the maintenance change, referred to as "Rationale in the Large" is stored in the Questions Options Criteria (QOC) notation (MacLean et al. 1989) and the rationale for the change implementation, or "Rationale in the Small" is captured as comments in the code that link the code to a folder ("Implementation Folder") for each maintenance request. The rationale is not in an argumentation structure but is a natural language description of the change. The implementation rationale is associated with the QOC option that it implements via a bi-directional traversal link.

14.5 Summary and Conclusions

A successful software system will spend the majority of its lifespan undergoing maintenance. As the development time recedes further into the past, the reasons behind the decisions that formed the product become increasingly inaccessible. This information, which includes the developers' initial intent, can be captured in the rationale so that it will be available when the software requires modification. In addition, capturing rationale for proposed changes can be used to compare that reasoning with the original requirements to ensure that consistency of goals is maintained.

In this chapter we described some major areas of software maintenance research and how the capture and use of rationale supports them. The costs and risks of maintenance are very high. The ability to obtain the deeper understanding of the software provided by the rationale is invaluable to assist with this process to minimize those costs and risks.

15 Rationale and Software Re-use

In this chapter, we describe how Software Engineering Rationale (SER) can be used during many types of software re-use, including how rationale can assist during Component-Based Software Engineering, with Software Product Lines and COTS-based software development.

15.1 Introduction

15.1.1 Software Re-use

Software re-use has long been promoted as a means to deliver software faster, cheaper, and with higher quality. While this is a worthwhile goal, its achievement is by no means guaranteed. There are many concerns that need to be addressed which vary depending on the type of re-use attempted and on each specific project that re-use is applied to.

There are many different types and meanings of software re-use. In some cases, entire systems are re-used, in others, segments of code. Re-use does not just apply to code—requirements, designs, documentation, test plans, test procedures, any development artefact could potentially be used in constructing a new system. The re-used artefacts can be developed within the company doing the reusing or can be developed externally and purchased for use. When re-using code, the code can be treated as a "black box" or modified to fit a new application. In this chapter, we will refer to the artefact being reused as the reused "item" where the item could be anything that could be reused in multiple software systems.

Re-use has the potential to significantly reduce cost and increase quality of software systems and to shorten the time to market for applications. There are some pitfalls though that could trap the unwary. Building new applications from existing code, components, or applications requires that these items be well tested and of high quality. If an application uses components developed by a third party then there is a risk that the third party may cease to support the component in the future or go out of business all together. There are also evolution issues if the re-used item evolves in a way that introduces incompatibilities. These potential problems should not discourage re-use but are risks that need to be considered when making decisions on when and how to use re-use in a development effort.

While re-use is often opportunistic, the most benefit can be derived from systematic re-use (Schmidt and Buschmann 2003). In systematic re-use, the re-use is intentional and the items re-used have been developed, tested and shown to be of high quality.

The determination of if, when, and how, to re-use software is a decision-making process and as such, benefits by the use of rationale when considering different re-use alternatives. This benefit continues as the software evolves when rationale can assist in determining if choices made earlier should be reconsidered and if earlier re-use alternatives might now be preferable.

15.1.2 Objectives of this Chapter

This chapter discusses different types of re-use and how rationale can support them. In particular, it focuses on four key types of re-use: patterns, Component Based Software Engineering, software product lines, and COTS-based software development.

15.2 Re-use: Concepts and Categories

There are many ways that software re-use types can be classified and categorized. Some re-use is opportunistic, such as copying segments of code from one application to another and modifying as necessary. Other re-use is planned from the start. The re-use granularity can be small, on the class or class library level, or large, by re-using entire applications as part of a larger system. In this section, we will describe some of the more common types of re-use.

15.2.1 Types of Re-use

Early empirical studies illustrated that the object-oriented paradigm supports software re-use (Lewis et al. 1991). This re-usability was taken one step further by the definition of Design Patterns (Gamma et al. 1995). Design Patterns are re-usable collection of classes that both capture solutions to common problems and that are aimed towards designing software in a way that will better support re-use and extension in the future by reducing, or isolating, dependencies between classes. The patterns serve as re-usable designs and instantiations of these patterns produce re-usable code. Some, such as the Iterator pattern, have been built into class libraries such as the C++ Standard Template Library. The concept of reusable patterns in software engineering has been extended beyond design and into process patterns (Coplein 1995; Ambler 1998), quality patterns (Houdeck and Kemper 1997), architecture patterns (Gomaa and Farrukh 1998), and more.

Another form of re-use is supported by Component Based Software Engineering (CBSE). The goal of CBSE is to develop software for less money and in less time by following a similar model to that in other engineering fields where new devices are composed of re-usable components often selected from a catalogue. For a software element to be considered a component it needs to conform to a component model, where the component model defines how components are composed into applications and how they communicate with each other. A component also needs to be deployed independently and composed into applications without requiring modification or customization. This composition needs to form to a composition standard (Councill and Heineman 2001). Example component models include the Common Object Request Broker Architecture (CORBA) (Object Management Group 2000; Wang et al. 2001), Microsoft's Component Object Model (COM) and its successor .NET (Ewald 2001), and Enterprise Java Beans (EJB) (Matena and Hapner 1999; Blevins 2001). Web Services can also be considered a type of component.

Product Line development is a form of re-use where a family of applications is developed from a code baseline. A software product line is "a set of software-intensive systems sharing a common, managed set of features that satisfy the specific needs of a particular market segment or mission and that are developed from a common set of core assets in a prescribed way." (Clements and Northrop 2002). These multiple product families could support different platforms (operating system, hardware), different operating environments (different peripherals), differences in functionality, or support different business processes (Sommerville 2007).

Another common form of re-use is in building products from commercial off-the-shelf software (COTS) or government off-the-shelf software (GOTS). Like with CBSE (which can be viewed as a form of COTS development), the goal is to reduce development time by utilizing software that is purchased, not built in-house. The success of these efforts depends on both the availability of suitable COTS systems to incorporate into the new system and in the flexibility of the requirements for the new system being developed. It is unlikely that a "perfect match" will be found between the COTS systems available and the requirements for the new application. Other challenges in COTS-based SE are the willingness to be dependent on the vendor(s) providing the COTS systems. Vendors can discontinue support (or just give poor support), change the API and introduce incompatibilities, increase pricing for new releases, and potentially cease to exist entirely.

15.2.2 Types of Rationale for Re-use

There are several types of rationale that can come into play when supporting software reuse. One is the *re-use candidate rationale*—rationale associated with the re-used artifact. The rationale behind the design and implementation decisions made when building the reusable item would give crucial insight into its functionality and quality. The rationale could point out what features of the item support reusability and also could point out what the overall quality goals were. This information can be invaluable in determining if the reusable item is suitable for a particular application. This information is also important as the reusable item evolves. If the design or implantation is changed in such a way as to "break" the API or make the application less reusable, that will affect anyone using the item who needs to stay consistent with upgrades and other new releases. It is especially important to capture any tradeoffs made between functionality and generality.

Another type of rationale that is useful is the *re-use approach rationale*—rationale for deciding how reuse should take place within a software application. There are advantages and disadvantages for reusing software versus building it "from scratch." It is important to capture these alternatives and decisions in case they need to be revisited again in future development iterations or later in system development if goals change. In some cases, the requirements of the new system will require adjustment in order to fit the services provided by a reused item. The reasons for these adjustments should be captured in the rationale for the requirements so that if issues arise because of the adjustments the developers will know why the adjustments were made.

Another type of rationale will be the *re-used component selection rationale*—rationale for selecting between alternative items for re-use. This is particularly applicable during CBSE and COTS-based development efforts. The rationale will give insight into the intent of the developers when making these decisions and can be used to re-evaluate choices as the systems (both the system being developed and the items being reused) evolve. The rationale will assist in keeping track of what information needs to be reviewed periodically as applications change and by providing information on alternatives that can be re-considered if the currently selected ones prove to no longer be feasible either due to lack of vendor support or changing requirements.

A final type of rationale is the *re-use dependency rationale*—rationale that captures the decisions made in development that are dependent on the selection of a specific reusable item. This information can be used later to assess the impact if that item can no longer be used in the system or if it has significant changes made to it that affect how it is used in the system.

15.2.3 Re-usable Rationale

Re-using code, documentation, designs, etc. promises to reduce the amount of effort required when developing software. In similar fashion, re-using the rationale for these items decreases the amount of effort required to capture rationale for the new system. Having a set of potential alternatives and the arguments for and against these alternatives will give developers a significant head-start when they need to make similar decisions in the future.

One methodology to support this is the use of Reusable Rationale Blocks (RRBs) (Hordijk and Wieringa 2006). RRBs are a collection of general design decisions, possible design alternatives for each decision, evaluation criteria, and ratings of each alternative based on that criteria. The collection of the RRBs forms a "generalized design space." When new problems need to be solved, "problem matching" is done against the set of RRBs to see if this is a decision that needed to be made in the past. If that is the case, the alternatives and criteria can be examined and adjusted, if needed, to fit the new problem. The RRBs are a form of reusable design knowledge and can help guide the designers toward solutions to design problems.

15.3 Applying Rationale

15.3.1 Rationale and Patterns

As mentioned earlier, there are many different types of patterns that can be used in software development. Patterns, a concept that initiated with Alexander's patterns for architecture (Alexander 1977), typically contain the description of the problem, a solution, or activities that comprise the solution, and the consequences, or results, of applying the pattern. These consequences can be viewed as the rationale for selecting the pattern.

Expressing the rationale inside a pattern description provides insight into when and how the pattern should be implemented (Bozheva and Gallo 2006).

The type of pattern best known in Software Engineering is the design pattern. While design patterns can be a good way to use knowledge of existing good designs to solve problems, they do need to be used with care. There is always the risk of building in flexibility and extensibility into a software product that does not necessarily need it. Flexibility and reusability only bring a cost and time savings if these capabilities are needed. Building software for easy extension and re-use does not come without cost so it is important to determine if this cost will be justified. If not, you run the risk of incurring the extra costs involved in carrying around a more complicated design (Beck 1999). The deliberation behind a decision on whether or not the flexibility/extensibility provided by a design pattern is needed can be captured in the rationale. The rationale can then be used to revisit these decisions if they need to be reconsidered at a later date.

If the decision to use a design pattern has been made, the rationale can be used to determine which pattern is most appropriate for a given problem. Gamma et al. (1995) describe several approaches for determining what design pattern is right for a given problem. Several of these approaches utilize different forms of rationale for the patterns. These include the intent for the pattern, given in unstructured text as part of the pattern catalogue, the applicability, which what problems a pattern applies to, and the consequences, which give some of the tradeoffs involved in choosing a pattern. The rationale for choosing a pattern should map to the causes of re-design that the pattern addresses.

The use of rationale to assist in pattern selection was the goal of the DRIMER (Design Recommendation and Intent Model Extended to Reusability) system (Peña-Mora and Vadhavkar 1997). DRIMER implemented the approach of "patterns-by-intent" where a design pattern is selected using the designers' intent and then code that implements that pattern is chosen and adapted, if necessary, based on the constraints for the system being built. Essentially, the design rationale for the system being developed is used to drive the pattern selection process and is used as an index into a repository of re-usable code.

One danger with using design patterns is the risk that future developers working with the code do not recognize the use of the pattern and why it is important for the development effort. Design Pattern Rationale Graphs (Baniassad et al. 2003) address this problem by representing the design patterns, and the rationale behind their use, and the source code that implements these patterns in a graphical format. The pattern graph is maps to the pattern description and the source code graph maps to the implementing source code. The developer, or maintainer, can explore both graphs to determine what the goals are behind the pattern implementation. The designer can explore the graphs using both regular expression based search and node expansion. The pattern graph captures alternative ways to implement each pattern and the rationale for each choice. During maintenance, using the DPRG assists the maintainer with exploring what design goals are relevant to the code that implements the design pattern.

Patterns do not just apply to the software artefacts. Patterns also appear in software process. Hagge et al. (2006) describe how process patterns exist for successful requirements engineering (RE) practices and how those patterns can be re-used to support process improvement. The process patterns themselves follow a structure that is very similar to rationale where the pattern description giving the problem, solution, context that guides it, and the experience that supports it maps to rationale where a question (decision) is posed, an option is considered, with arguments given for and against it that involve evaluation criteria. A collection of RE patterns is being saved in the Requirements Engineering Patterns Repository (REPARE).

Process knowledge is also distributed in pattern form in Agile Patterns (Bozheva and Gallo 2006). Agile patterns describe alternative ways to address practices, concepts, and principles encountered and utilized when applying agile methods. As in RE patterns, the rationale is part of the pattern description, although the decision criteria do need some further definition. These patterns can be used as guidelines for activity selection, as a means for supporting knowledge transfer by providing "past knowledge" from the software engineers who solved the problem expressed by the pattern in earlier development efforts.

15.3.2 Rationale and Component-Based Software Engineering

Rationale could be captured and used in making a number of important decisions in CBSE. For component providers, these include the component model(s) to support and the granularity of the components. For component consumers, the decisions include when components should be used, which components to use and which component model best suits the application. Rationale can both support making these decisions and document their results.

When working with COTS components, the initial difficulty faced is identifying candidate components. There are a number of possible methods for finding components. The ideal approach would be for all components to provide a common description model. While the internet can be used to access multiple component catalogues, the different collections of components focus on different types of components or different aspects when describing them (Requile-Romanczuk et al. 2005). The search is also made difficult when designers may not be able to specify precisely what they need. This requires flexible retrieval mechanisms to work with the developer in formulating their requests (Fischer et al. 1991).

It is also important to understand what relationship components have with each other. Are there component characteristics that are likely to impact each other? Are there dependencies between components? These factors are especially important for dynamically configured components, such as Web Services. It would be valuable to have a way to reason over these services to determine how they will impact system design (Gannod et al. 2007).

While components are selected based on the functionality they provide, there are other characteristics of the component that can influence its selection. The Unified Specification of Components framework (UnSCom) (Overhage 2004) extends the concept of design by contract to CBSE. The composition contracts are specified on multiple contract levels to define the component interfaces. These levels include both the functionality of the component and information about component quality.

15.3.3 Rationale and Software Product Lines

Software product lines utilize re-use by creating closely related applications as application families where portions of the application that are the same are shared. Application families are described by commonality and by discriminants where a discriminant is a requirement that differentiates between systems (Mannion et al. 1999). These discriminants serve as decision points where choosing different alternatives results in different products. The Method for Requirements Authoring and Management (MRAM) uses a requirements metamodel to describe the requirements that comprise the application family. Several of the attributes that describe each requirement can also serve as its rationale. These include stability, verifiability, complexity, cost, staff-knowledge, and technology. This information is available to stakeholders who can then look at the impact of selecting different requirements at the choice points indicated by the discriminants. MRAM and its supporting meta-model are used by the Tool for Requirements Authoring and Management (TRAM) to use a set of application family requirements to select those for a single product and generate a system model.

One challenge in Product Line Engineering is deciding how the product line should be structured. Planning the product line is also known as product line scoping. The core task of this process is looking at which functionality in the product line will have the best return on investment when re-used (Schmid 2002). This process involves many activities that can best be captured in the rationale for the product line design. The planners need to investigate different scoping alternatives and their advantages and disadvantages. This includes looking at tradeoffs between business objectives and evaluation of risks (Schmid 2002). The rationale can serve as a basis for negotiation and as input to any cost-benefit analysis that may be required. The rationale can capture the relative importance of each evaluation criteria. When the decisions are made, rationale should also be collected to indicate which portions of each product are meant to be re-usable and which are not.

Knodel and Muthig (2006) developed a process to capture architecture decisions and their rationale. They focused on these decisions because decisions made when developing the architecture for the product lines are especially important because of their strategic value to the organization. Capturing the key decisions that drive the architecture and the rationale behind them serves several important goals. The process of capturing and discussing the rationale provides a mechanism for identifying and documenting what the important criteria and issues are behind the product line architecture. This includes prioritizing these criteria and using these priorities to evaluate the candidate design alternatives. This process also supports the negotiation performed by the architects and other stakeholders. The resulting rationale can be used later to defend these decisions to interested parties who were not actively involved in the decision-making and can also be invaluable to any new developers who need to learn about the architecture.

Another decision that needs to be made when designing a software product line is how it will be configured and managed. Architectures can be configured at two points in the development process: deployment-time configuration and design-time configuration (Sommerville 2007). Deployment-time configuration means that the system can be configured for a specific customer using configuration files. Design-time configuration works with the core functionality of the product line but includes new or modified components in order to support specific product needs. Rationale can be used to document the advantages and disadvantage of each approach and the reasons for making this decision. It can also be used to document how the different product line components meet the reconfiguration goals to ensure that evolutions to the product line do not restrict reconfigurability.

15.3.4 Rationale and COTS-Based Software Engineering

The success of a COTS-based project will depend on a number of key decisions which should be documented in the rationale. The first decision is if COTS products should be used at all. Fifteen COTS-based projects in a NASA environment were studied by Morisio et al. (2000). The goal was to capture the actual process and to identify what the differences were between COTS-based and actual development. This study looked at the decision that determined when COTS-based development is appropriate. There are tradeoffs that need to be made when determining if it makes more sense to buy or to build. These include cost, risk, and requirements. The requirements need to be flexible enough to accommodate some adjustment in order to conform to COTS systems available. Morisio, et al. recommend that the requirements be sketched out initially with only enough detail present to choose COTS products to incorporate. This needs to be done in view of the danger that Kontio (1996) pointed out where "fuzzy" initial requirements may lead the decision-making process to focus on easier to investigate technical issues that may not be as important as the application requirements when making the COTS selection.

The other key decision (which is not independent of the first) is which COTS products should be used. The ability of the system to meet the application requirements is important but there are also many non-functional requirements that play a role in the project's success. The decision-making process will involve investigating the licensing cost of the product (this can be a significant factor if the product being developed will be installed at multiple locations or purchased by multiple customers) and the maturity of the vendor and product. Determining the Technology Readiness Level (TRL) for each system incorporated is mandated when developing systems for the United States Department of Defense (DoD 2002). Capturing this and other non-functional criteria in the rationale both documents the selection process and makes it easier to reassess decisions if criteria or evaluations change over time.

The OTSO method (Kontio 1996) was developed to support a COTS selection process that was more systematic than the ad hoc methods that are often used. The method defined what tasks had to occur in the selection process, a hierarchy of evaluation criteria, and a model of costs and value for the COTS alternatives. Each alternative was given values for the evaluation criteria and then ranked. The Analytic Hierarchy Process (AHP) (Saaty 1990) was used to rank the alternatives. In this process, alternatives are compared in pairs, rather than given some absolute evaluation value.

The evaluation criteria and their value for each alternative form the rationale for the COTS alternative selection.

Rationale can also be captured for the process followed to make COTS technology selections. The Resources-based Approach for COTS Evaluation and selection (RACE) uses a process model that contains activities that can take place in COTS evaluation and selection (Mohamed et al. 2005). The selection process can be customized based on the project domain. The domain characteristics would form the rationale for process choices.

15.4. Summary and Conclusions

Software re-use has become an integral part of many, if not most, software development projects. While the potential cost savings are considerable, there are also considerable risks towards depending on software delivered elsewhere. It can also be difficult to identify and select software to re-use and to integrate the re-used items into a new system.

In this chapter, we have presented several common types of re-use and how rationale could assist in making these efforts more successful. Re-use is something that should be pursued when possible but involves many decisions that require considerable thought. The ability to document those decisions in a structured way through the rationale for the system helps to assure that the choices made are well justified. The presence of the rationale also provides invaluable insight to future developers maintaining or re-using the system.

Part 4
Frameworks for Rationale-Based Software Engineering

The case in support of rationale is a compelling one—the ability to capture and encode the decision-makers' intent as part of a knowledge management strategy aimed at using this knowledge to assist with future decisions so that we can learn from the past, rather than repeating it (or repeat it only when past decisions were successful). The importance of rationale and its potential value has resulted in a significant amount of research over the past thirty years yet there still remain many obstacles towards its acceptance and use in practice. Still, advances in technology have resulted in new opportunities for integrating rationale into practice and the increasing awareness of the relationship between process and product quality suggest that the reluctance to invest up front effort for later benefit may be lessening.

The challenge is to move rationale outside the laboratory and into practice. Studies have shown that it takes 15-20 years to mature a technology (Redwine and Riddle, 1985). In order to successfully transition a technology into practice it is important to understand what both the *obstacles* and *benefits* of that technology are.

In order to build a Rationale Management System (RMS) to support RBSE, we need to identify where and how rationale can be used in software development (benefits) and capture these uses, along with concepts needed to compare rationale approaches and relate them to software engineering, in a Conceptual Framework (Chapter 16). We also need to develop an Architectural Framework (Chapter 17) that identifies issues (obstacles) that must be addressed by an RMS architecture in order to successfully support software engineering. Past work in rationale has indicated that it shows great promise in providing significant benefits to software development and we need to look ahead (Chapter 18) to determine how those benefits can best be disseminated into software development approaches, processes, and tools.

16 A Conceptual Framework

Exploiting the full potential of rationale in software engineering requires a comprehensive understanding of that potential. Such understanding must be based on a conceptual framework that describes how and where rationale usage can support SE. This framework should identify where and how rationale can be used in software projects. It should also provide a set of concepts for comparing proposed approaches to rationale and for relating them to the various aspects of software engineering.

16.1 Introduction

16.1.1 What a Conceptual Framework Should Do

Understanding the full value of rationale in software engineering (SE) requires a conceptual framework that enables description of the ways in which rationale can support SE. As used here, the term *conceptual framework* means of a set of ideas and terms for describing the problems in an application domain and the means for solving them. A conceptual framework for rationale usage in the domain of software engineering must do three things: 1) provide connections between concepts of rationale and concepts of SE to enable description of how rationale can support or fail to support SE, 2) identify how different rationale approaches differ with respect to the goals of SE and 3) identify the goals and success criteria for rationale usage in SE.

What is needed is a common and unified framework that can relate all major rationale approaches to SE. This framework should not be expected to settle the many disputes among the proponents of different approaches, but it should clarify these disputes by revealing differences and commonalities amongst rationale approaches. To do this, it must include terms and ideas that reflect significant similarities and distinctions but exclude those that do not.

16.1.2 Objectives of this Chapter

The overall objective of this chapter is to describe a conceptual framework that can serve as a basis both for practical use of rationale in SE and for research on such use. To accomplish this, the chapter will attempt to describe the essential facts, criteria and concepts of rationale usage in SE. It will provide a terminology that organizes these entities into a unified framework. The framework provided here will not attempt to cover all concepts of rationale and software engineering. Instead it will focus on just those needed for *using rationale to support SE*. This will provide a foundation for methods and software tools that support rationale usage in SE.

The discussion begins with a section describing the general goals of rationale usage in SE. The next section groups proposed rationale usage into types of approaches, specific approaches within each type, and specific methods of usage within each approach. The section after this explores the range of use of the *decision-centric type of approach* in SE. The section following this explores use of the *usage-centric type of approach*. Following this is a section on rationale in iterative software development, then a section of challenges to rationale usage. The chapter concludes with a brief summary of the intent and contents of the chapter.

16.2 General Goals of Rationale Usage in Software Engineering

The overall goal of rationale usage in SE is simply to help software engineers achieve their goals. Typically this means helping the developers, maintenance personnel, users and other stakeholders to achieve their goals as well. The goal of *rationale-based software engineering* (RBSE), which is the theme of this book, is to increase the usefulness, usability and use of rationale in SE.

Ultimately, the way in which any kind of rationale aids SE is by helping to improve the quality of its decision making. In this sense all rationale usage is *prescriptive* in that it assumes that the quality of decision making in SE is not as good as it could be and that rationale usage can make it better. Even when rationale is recorded without any intention of influencing decision making, it can do so by helping software engineers to remember what has been decided and why or by serving as an aid to future projects.

There are two basic ways in which advocates of rationale usage seek to improve decision making. One is by providing better information for decision makers. The other is by prescribing decision making processes aimed at eliminating flaws in reasoning and failures to take important information into account. Advocates of rationale usage typically claim that it can help to make decision making better informed, more correct, more consistent and/or more complete.

While the ultimate goals of rationale usage are prescriptive, this does not necessarily mean that they are prescriptive with respect to the decisions for which rationale is captured. Sometimes they are merely *descriptive* in the sense that they seek only to describe, i.e. document, the reasoning behind these decisions without influencing it. In such cases, the documented rationale is used to inform other decisions—for which it is prescriptive. For example, the rationale of designers might be captured without influencing their decision making and then used to influence the decisions of people constructing the artifact. In fact, this approach is common.

16.3 Rationale: Types of Approaches, Specific Approaches and Methods

There are two fundamentally different types of rationale approaches. One focuses on *decision making by SE personnel*, the other on *the experiences of users*. Our conceptual framework labels the former type *decision-centric* and the latter *usage-centric*. Decision-centric approaches model the reasoning of people involved in decision making, including what decision tasks are undertaken, what decision alternatives are considered, how these alternatives are evaluated and how different decisions are related. Usage-centric approaches model the reasoning of people about an artifact, such as software, based on their attempts to use it. This reasoning centers on evaluations of artifact features as they are experienced in the context of use.

Within each type of approach there can be one or more specific approaches. For example, both Issue Based Information Systems (IBIS) (Kunz and Rittel 1970) and Questions Options and Criteria (QOC) (MacLean et al. 1989) are specific approaches within the decision-centric type of approach to rationale. Similarly, Scenario-Claims Analysis (SCA) is an approach within the usage-centric type of approach.

There may be several distinct methods of using rationale within a given rationale approach. For example, there are several methods of using IBIS. While such methods share commitments to core IBIS concepts, such as the basic IBIS rationale schema, they may differ on the details of the schema. Thus, for example, Conklin and his colleagues use the IBIS approach (Conklin and Begeman 1988) (Conklin and Burgess-Yakemovic 1996), but some of the details of their schema differ from Rittel's.

Another way rationale methods can differ is in whether they are 1) process-oriented or 2) structure-oriented (Lee and Lai 1996). Process-oriented methods stick closely to the temporal sequence in which rationale statements are actually generated; they thus provide a history of the reasoning process. Structure-oriented methods, on the other hand, abandon the temporal order of reasoning in favor of its "logical" structure. They thus provide a sort of idealized model of how ideas relate in rationale. Conklin and Burgess-Yakemovic (1996) use IBIS in a purely a process-oriented manner, but McCall (McCall 1979b) has shown that IBIS can also be used in a structure-oriented manner. Other decision-centric approaches also can differ in their method of use. For example, QOC is strictly structure-oriented, but DRL and PHI can both be used in either process- or structure-oriented methods.

16.4 Decision-centric Rationale in Software Engineering

There are two, closely related concepts that provide the most obvious points of connection between rationale research and SE. These are the concepts of *decision* and *decision making*. Most of the rationale methods described in this book deal explicitly with decision making; and it seems intuitively clear that software engineering involves the making of many decisions. The section below provides evidence for this intuition by looking more closely at the nature of decision making in rationale approaches and in SE.

16.4.1 Decision Making in Rationale Approaches

16.4.1.1 Decision Making as Question Answering

The rationale approaches that are relevant here are the decision-centric approaches, six of which were introduced earlier in this book: IBIS (Issue Based Information System), PHI (Procedural Hierarchy of Issues) (McCall 1991), QOC (Questions, Options and Criteria), Potts-Bruns (Potts and Bruns 1988), DRL (Decision Representation Language) (Lee 1991) and RATSpeak (Burge and Brown 2006). In all these methods 1) decisions are represented explicitly and 2) all other rationale exists entirely for the purpose of helping to make these decisions. To understand whether and where rationale approaches can support decision making in SE, we need first to understand how these approaches represent decision making and then to look at the various aspects of SE to see what decisions might be represented in this manner.

What counts as being a decision in decision-centric rationale? While the answer to this question depends to some degree on which rationale approach we are talking about, there is nevertheless a criterion that is common to all of them. *For something to be counted as a decision in these methods, it must be possible to state it as a question to be answered.* In fact, all approaches except DRL represent decisions exclusively as questions. While DRL sometimes uses other representations for decisions, its inventor (Lee) has declared that DRL decision tasks are the same entities represented as questions in QOC and IBIS (Lee and Lai 1996). From this we can infer that decision tasks in DRL also correspond to the question-based decision tasks in the Potts-Bruns approach and RATSpeak. This means that any decision task that is represented in DRL can be paraphrased as a question. So, for a decision task to be dealt with by any of the decision-centric rationale methods, it must be capable of being represented as a question. This provides a convenient means for identifying the parts of SE that are candidates for use of decision-centric rationale methods.

What needs to be done, then, is to identify the parts of SE that can be represented as questions to be answered. Once we have done this, we can take the next step and see whether the types of questions and question answering processes featured SE can be matched with the kinds of questions and question answering processes dealt with in different rationale approaches.

16.4.1.2 Question Answering through Deliberation

The common thread in all decision-centric approaches to rationale is that decisions are made, i.e. questions are answered, through a process of *deliberation*. The transitive verb *to deliberate* is defined by The Random House Dictionary of the English Language (Second Edition, Unabridged) as follows: "to weigh in the mind; consider: *to deliberate a question*." But a more detailed definition is needed here. This is accomplished by accepting this dictionary definition but defining "*to deliberate a question*" as follows: to evaluate one or more proposed answers to a question. Deliberation in this more precise sense is common to all decision-centric approaches to rationale.

Almost all decision-centric approaches use *argumentation* to evaluate proposed answers. One notable exception is the problem-centered approach of Lewis, Rieman and Bell (1996). This approach evaluates proposed answers not by arguing their merits but by testing them using a suite of problems.

There are two major types of approaches to argumentative evaluation. One allows arguments for and against the proposed answers as well as arguments for and against other arguments in a multilevel, directed acyclic graph (DAG) structure of arguments. This is the approach used by IBIS and PHI, for example. The other approach to argumentative evaluation differentiates between the structure of the arguments on proposed answers and the structure of arguments on other arguments. In particular, argumentation on proposed answers uses *criterion-based evaluation*, which consists of 1) the statement of a criterion, e.g. a goal, and 2) an assessment of the proposed answers with respect to the stated criterion, these two elements in effect constituting a single argument for or against the proposed answer. Arguments for and against other arguments are in the same basic form as in the other approach to argumentative evaluation, i.e. a DAG of arguments. QOC and DRL, for example, both use this second approach.

There is one additional aspect of deliberation that is allowed by some but not all rationale approaches, and that is *dependency on the outcome of other decision making, i.e. question answering.* PHI, DRL and RATspeak allow this. In fact, PHI's overall structure of decisions is based entirely on such relationships. DRL allows dependencies in the form of several types of relationships between decisions. RATspeak also provides a special type of argument dedicated to representing dependency relationships between the answers proposed to different questions. Such an argument shows more specifically than PHI or DRL how one decision can depend on another. (See Section 16.6 below for a more detailed account of dependency relationships between decisions.)

16.4.2 Question Answering in Software Engineering

16.4.2.1 Questions in Software Engineering

The next thing to do is to list the various *aspects of SE and see to what extent they can be viewed as involving question answering processes.* There are two categories of such aspects. One category includes types of software-related *activities.* The following list of basic activities is adapted primarily from SWEBOK (the Software Engineering Body of Knowledge) (Software Engineering Coordinating Committee 2004). Though this document is by no means universally agreed upon, there is nevertheless broad consensus that SE involves the activities listed here:

- *Project inception—i.e. determining that a new or revised system is needed*
- *Requirements engineering*
- *Design and re-design*
- *Construction—i.e. implementation in code*
- *Testing*
- *Use*
- *Maintenance*
- *Configuration management*

The second category lists aspects of SE that deal with *relationships among various activities and stakeholders* in a software project:

- *Coordination within the SE team*
- *Collaboration amongst members of the SE team*
- *Participation of users in development*
- *Feedback and feedforward between different SE activities*
- *Management of the overall SE effort for a project*

These lists could, of course, be further elaborated to several more levels of detail. But the current level is adequate for demonstrating that there are many potential aspects of SE where rationale can find application.

To show that rationale has potential use in the above-listed aspects of software engineering, it is sufficient to show that there are questions that these aspects seek to answer through reasoning. Examples of such questions for each of the above-listed aspects are shown below. The list below only includes representative examples of the many questions that the aspects might deal with. The goal here is not to be comprehensive but merely to show that there are many candidates for rationale usage across a wide spectrum of SE aspects.

- **Software-related activities**

  - *Project inception—i.e. determining that a new or revised system is needed*
    - Why do we need a new or renewed system?
    - What is the purpose of the system from the perspective of its stakeholders?
    - What is the description of the user organization and the work that the users will perform for this organization using the software?

- How feasible would the software be technically and economically?

o *Requirements engineering*
  - What functions should the software fulfill in terms of users interacting with the system?
  - What are the functional and nonfunctional requirements of the system?
  - What are the requirements for the technologies to be used for constructing and operating the software?

o *Design and redesign*
  - What is the design of the software architecture to be—i.e. how the software is organized as subsystems?
  - What are the specific behaviors of these subsystems to be?
  - What is the design of each of these subsystems to be?

o *Construction*
  - What platform and coding technologies should be used to construct the system?
  - What public and in-house standards should be adhered to in construction?
  - How should the software be constructed to facilitate verification?
  - How should the software be constructed to facilitate change?

o *Testing*
  - What are the defects and problems with the constructed software?
  - To what extent do the implemented design features satisfy the stated requirements? (verification)
  - To what extent does the code successfully implement the design features? (conformance testing)
  - To what extent do the implemented design features satisfy user expectations? (validation)
  - How should the subsystems of the software architecture be tested?
  - How should the integration of the subsystems be tested?

- What test cases should be used in view of the given limitations in resources and schedule?
- Which test techniques should be used?
- How can the time between the creation of errors and their detection through testing be minimized?

   o *Maintenance*
- How can this software best be maintained?
- What enhancements are needed in the software?
- What problems are users having with the software?
- How can needed modifications to the software be made without breaking existing functionality or degrading performance of the system?
- What should the plan be for maintenance of this software?
- What modifications are needed to keep the software usable in a changing environment?
- What needs to be done to avoid potential future faults?
- What needs to be done to reduce the complexity of this evolving software?

   o *Configuration management*
- What is the plan to be for software configuration management (SCM)?
- What organizations should be involved in the SCM process?
- Which organizational entities should be responsible for which SCM tasks?
- What are the necessary sequences of the SCM tasks?
- What are the relationships of the SCM tasks to the project schedule and milestones?
- What tools should be used to support the different SCM tasks?
- How should the SCM plan be implemented?

- **Relationships among stakeholders and activities**

   o *Coordination within the SE team*
- What are the potential conflicts between the decisions and decision criteria used by different members of the development team?

- What are the dependencies between the activities of the various team members?
- How can team members be kept aware of changes to decisions about requirements, the design and the implementation of the software?
- How can team members be made aware of the effects of changes on their work?
  - o *Collaboration amongst members of the SE team*
    - What are the potential conflicts between the decisions and decision criteria used by different members of the development team?
    - What is the untapped potential for the work of team members to support the work of other team members?
    - Which members of the SE have knowledge that would be useful to other members of the team?
  - o *Participation of users in development*
    - What are user reactions to proposed features of the software?
    - What are user reactions to implemented features of the software?
    - What can be done to motivate users to participate in the development of the software?
    - How can users be made aware of the effects of proposed changes on their use of the software?
  - o *Feedback and feedforward between different activities*
    - Does the design of the system satisfy the requirements?
    - Does the constructed software correctly implement the intended design?
    - To what extent does the implemented software satisfy or fail to satisfy actual user needs?
    - Can the architecture of the system adapt to changes in requirements?
    - Does the design of the software facilitate its construction?
    - Does the design of the software facilitate its use?
    - Does the design of the software facilitate its maintenance?
    - What does the plan for testing imply for the design of the system?

- To what extent do the results of testing reveal that the design of the software has been effective in preventing the occurrence of errors?

o *Management of the overall SE effort for a project*
- What should nature and structure of the project tasks be?
- What resources should be allocated to which tasks?
- Which software quality management processes should be utilized?
- What software lifecycle model should be used?
- What lifecycle processes should be selected for the project?
- What software methods and tools should be used?
- What should the organizational structure be for the project?

Few would argue that answering the questions shown above requires anything less than careful and informed reasoning. As a consequence, each question represents an opportunity for both rationale capture and delivery. Each is also a potential candidate for the use of prescriptive rationale methods designed to improve the thoroughness, consistency and correctness of reasoning.

This list of questions shows that rationale usage is in no way limited to design. In fact, it suggests that design rationale constitutes only a small minority of the potential types of rationale in SE. It is precisely for this reason that this book employs the term *software engineering rationale* (SER) instead of the traditional term *design rationale* (DR) as the umbrella term for research and applications of rationale in SE.

The above-given list of questions is really just the tip of the iceberg. Each of the questions listed can lead to many other questions at lower level of detail that would need to be answered. And, of course, in addition to the questions shown above, there are many other questions at the same level of detail that are not shown here.

16.4.3 Using Decision-centric Rationale in the Full Spectrum of SER

Being able to state a decision as a question is *necessary* for applying decision-centric rationale methods, but is it *sufficient*? Answering this question requires knowing whether there are any obstacles for application of a rationale approach or method to particular decision making tasks. For rationale approaches in general, the only additional condition is that the decisions not be based on pure, inexplicable intuition rather than explicit reasoning. There are also two potential problems that can prevent use of particular approaches or methods of rationale usage:

1. a given rationale approach or method is intrinsically applicable only to certain types of decisions
2. the way decision making is accomplished in SE is incompatible with the way decision making is represented or accomplished in a given rationale approach or method

The following sections look at each of these problems.

16.4.3.1 Rationale Approach Restricted to Certain Types of Decisions

IBIS and its PHI derivative have no restrictions as to what kinds of decisions they can deal with. While Rittel originally intended IBIS only for use with *controversial* questions, i.e. those which stakeholders disagreed about, this restriction has long since been abandoned by most users of IBIS, including McCall (1979a, 1986), Conklin and his colleagues (Conklin and Begeman 1988; Conklin and Burgess-Yakemovic 1996), Buckingham Shum and his colleagues (Buckingham Shum et al. 2006).

QOC is restricted by its authors to use only for *design space* questions—i.e. questions denoting decisions on the features that an artifact should have (MacLean et al. 1996). This presents an apparent obstacle to use of QOC for the full range of decisions in SE. It should be noted however, that Dutoit and Paech (2000) have shown that QOC can be refined to be applicable to requirements engineering. Despite the intentions of its inventors, QOC might turn out to be usable for other many other types of SE decisions as well; but to date, this has not been demonstrated. The only feature of QOC that offers any possibility of limiting its range of application is its method of evaluation. QOC has a requirement that decision alternatives—called *options* in the method—be evaluated against explicitly stated criteria, and it is unclear whether this requirement can be satisfied for all decisions in SE.

This issue arises again in the case of DRL, because its treatment of decisions corresponds closely to QOC's—especially in the use of criterion-based evaluation of decision alternatives. While Lee never states that DRL is restricted to the *design space* decisions that QOC focuses on, the various examples given for DRL (Lee 1990; Lee 1991; Lee and Lai 1996) deal only with such decisions. This, however, might merely represent the exclusive focus on design rationale that characterized the early literature on rationale.

It is important to note that in addition to *decision problems, DRL* also allows *question-answering* in its schema, though it provides no schema for question answering other than the element types *question* and *claim* plus the relationships *answers* between them. If the entities that DRL labels *decision problems* do not in fact represent anything more that the *design space questions* found in QOC, it is still possible that DRL could still be used for other types of SE decisions by augmenting its question-answering schema to allow IBIS-type multi-level argumentation. One simple way to represent this argumentation would be to use DRL's *claims* linked by its *supports* and *denies* relationships. This is a very minor extension of DRL's current schema.

16.4.3.2 Decision Making Processes in SE Incompatible with Decision Making Processes in Rationale

The concept of decision making as defined for decision-centric rationale— i.e. as question answering—has shown that many decisions in SE are candidates for rationale usage. But these are candidates only. To understand whether rationale methods can be applied to them requires understanding whether the *processes of decision making* as they are represented in rationale approaches and methods is compatible to the processes of decision making in SE. In order to answer this question with respect to any given rationale method it is, first of all, important to know whether the way in which a rationale approach is to be applied is *descriptive* or *prescriptive* with respect to how that decision is made. Prescriptive use of a rationale approach dictates certain processes that must be followed or certain information that must be used in making the decision. Descriptive use, by contrast, makes no attempt to impose rules about decision making processes or what information must be used. Instead, it merely documents whatever discussion happens to arise, by categorizing statements according to its schema—e.g. as issues, positions, arguments, and so forth in the case of IBIS.

A good example illustrating the prescriptive/descriptive distinction is the IBIS approach. As Rittel originally intended that it be used, IBIS was strongly prescriptive in the sense that he sought to change the way in which decisions were made. In particular, Rittel used IBIS to promote the idea of decision making as being based on debate of decision alternatives amongst a wide spectrum of stakeholders. As a consequence, the generation of alternative positions to each issue was mandated, as was the generation of arguments for and against both the positions and other arguments.

But as time progressed, IBIS was frequently used descriptively as well. In such cases, it *could* record alternative positions as well as arguments for and against positions and arguments *if they happened naturally to arise*; but it was not used to direct the discussion.

A prescriptive version of IBIS, or any other rationale approach, might well be in conflict with certain approaches and methods to software engineering. So, for example, if a software method refused to recognize differences of opinion as legitimate among reasonable and informed people, the prescriptive version of IBIS would clash with it. Or if, as in Boehm's Spiral method, the emphasis is on resolving differences of opinion quickly and smoothly—as is the case with the WinWin rationale method (Boehm and Kitapci 2006)—then Rittel's original goal of using IBIS to fan the flames of debate among stakeholders (Rittel 1972) might well be seen as counterproductive to the WinWin goal of showing how all stakeholders can be winners. Prescriptive use of any given rationale method *might* well conflict with software engineering methods, but whether they *actually* conflict depends on which rationale method is used and which software method is used. Whether there is in fact a conflict must be worked out on a case-by-case basis that compares a given rationale method to a given software method. Unfortunately, such a case-by-case comparison is beyond the scope of this book.

Where rationale methods are used in a purely descriptive manner, there can arise no conflict with SE methods. It should be noted, however, that rationale methods can be used descriptively with respect to some activities and prescriptively with respect to others. A common example of this is when rationale methods are used descriptively for *rationale capture* for one SE aspect—e.g. design—and prescriptively for *rationale delivery* with respect to another SE aspect—e.g. project management. Managers may want to monitor the activities of designers in a way that does not dictate what the designer do, while at the same time using the information for management tasks such as coordinating the work of others with the work of the designers. This use of rationale is prescriptive with respect to managers in the weak sense that it *informs decisions that they make*. Ultimately, rationale is of no value if it is not prescriptive in the sense of influencing some SE decision making in a current or future project. Keeping track of these influences is an important aspect of rationale usage in SE.

As conceived of by its inventors, QOC is purely descriptive and so presents no possibility of a clash with any SE method. DRL's author makes no assertion about whether it should be used prescriptively or descriptively; so nothing prevents it from being used in a purely descriptive manner. While both IBIS and PHI were originally intended to be prescriptive, there have since been many uses of both that are purely descriptive.

Without attempting to enumerate all software methods and how they might conflict with prescriptive uses of rationale methods, we can nevertheless describe the types of conflict that *could* arise. This can be done by showing the specific prescriptive assumptions of the prescriptive rationale methods. The only two rationale methods that have prescriptive modes about the processes by which decisions are made are IBIS and PHI.

Many current users of IBIS use it in prescriptive mode. This includes the "process-oriented approach" to IBIS advocated by Conklin and Burgess-Yakemovic (1996) and the IBIS work of Buckingham Shum et al. (2006) with the Compendium hypertext system. Both uses of IBIS advocate a *constructive disruption* of decision making processes that according to Buckingham Shum et al. (1997) is based on the notion that, "deeper understanding of a domain comes through the *discipline* of expressing knowledge within a structural framework, working to articulate important distinctions and relationships." The compatibility of these ways of using IBIS with various SE methods is largely an open question.

PHI originally came with a requirement to use a top-down, breadth-first approach to raising issues; and this might well conflict with certain SE methods. But this procedural prescription has been abandoned in later uses of PHI.

16.5 Usage-centric Rationale in Software Engineering

The primary measure of software quality is its value to its users. Therefore, an important complement to decision-centric rationale is usage-centric rationale, which documents the evaluation of a system by its users on the basis of their experiences in using the system. Perhaps the best way to document the experience of a user is in terms of a usage scenario, i.e. a history of the sequence of steps involved in usage of a system. Carroll and Rosson have pioneered this type of rationale documentation with the Scenario-Claims Analysis (SCA) approach.

Understanding the full potential of SCA as a tool of SE requires answering two questions: Where can SCA capture rationale within the overall SE process? and Where can SCA rationale be used within the SE process? The obvious answer to the first question is that SCA always captures rationale during use of a system, but this answer can be misleading if the term *use* is understood too narrowly. While SCA can capture rationale during *actual use of a fully designed, constructed and deployed system*, this is by no means the only type of use that is relevant. It can also capture rationale during use of 1) not-yet-deployed systems, 2) partially constructed systems, 3) design prototypes and 4) prototypes created merely to elicit requirements. It can even capture rationale from simulated use of not-yet-constructed designs.

Where can captured SCA rationale be used within the SE process? In other words, what can SCA rationale be used *for*? SCA is above-all evaluation tool; it therefore has three main types of uses: 1) to rate a single system from the perspective of its users, 2) to determine the best of several competing systems (or subsystems) and 3) to provide feedback about use to the SE activities of requirements determination, design, construction and maintenance of the system. This feedback, of course, is aimed at informing and motivating the next iteration of each of these activities. SCA is thus a crucial driver of change and iteration in these activities.

The main relationship of SCA to the decision-centric approaches is that its rationale should be fed back into and become a part of the rationale on decision making. SCA's evaluation goes beyond the evaluation used in decision-centric rationale in several crucial respects. SCA does not merely evaluate individual decisions; it evaluates the *collection of decisions that constitute a design*. And it does this from the consistent perspective of a user engaged in a usage scenario. SCA's evaluation of a design can only begin after many decisions have been made by the designers. It thus constitutes an empirical *test* of those decisions.

16.6 Rationale and Iterative Software Development

Chapter 2, entitled "What Makes Software Different," pointed out that software development can be done using iterative processes that are not feasible in the development of many other types of artifacts. The term *iterative software development* is used here to refer to the repeated construction and use of preliminary versions of software to obtain feedback that informs requirements determination and design. As used here, this term encompasses such labels as *evolutionary* (Rajlich 2006), *incremental* (Larman 2003), and *agile development* (Larman 2004) as well as *Extreme Programming* (Beck and Andres 2005). While iterative software development is not performed on all development projects, it continues to grow in popularity as a means to address the often volatile nature of software requirements. It is important, therefore, to ask what the implications of iterative development are for rationale.

16.6.1 A Rationale-Based Account of Iterative Development

Instead of thinking of rationale merely as an add-on to iterative development processes, it is useful to try to understand iterative development as itself being a rationale-based and rationale-driven process. In fact, the motivation for and nature of iterative development can be explained by the nature of the reasoning processes that underlie the creation of software. While a complete rationale-based account of iterative development is beyond the scope of this chapter, the paragraphs below provide a sketch how such an account can be constructed for some aspects of iterative development.

Motivation for iterative software development arises in large part from the inherent inadequacy of the rationale for development decisions when those decisions are first made. This inadequacy takes the form of both incorrectness and incompleteness. Perhaps the most basic way in which the rationale is incomplete or incorrect is in its listing of user requirements, which are the grounds for much of a system's rationale. One reason for this is that software development takes time, and during this time user requirements can change. Another reason is that users themselves do not have explicit knowledge in advance of their own requirements. There is, for example, the IKIWISI (I'll know it when I see it) effect that happens when software is highly interactive. The effect exists because users cannot anticipate the results of these interactions and thus cannot predict their needs prior to these interactions. Yet another reason for the failure to identify the requirements in advance is that the satisfaction of some requirements can cause others to surface, because it changes the work environment or the priority of values.

Another reason for the incompleteness of rationale is that this rationale for decisions is based largely on the desirability or undesirability of the consequences of proposed decision alternatives. The problem is that there are important consequences that *cannot be foreseen at the time a decision is initially made.* They only become known later. This phenomenon is sometimes referred to as "The Law of Unintended Consequences." Once discovered, these consequences can motivate developers to revise development decisions.

Unforeseen consequences include some of a decision's impacts on subsequent SE activities—for example, impacts of a design decision on the construction and use of the software. While many of these impacts are intended, some might not be. A decision to include something as a design feature has the intended consequence of the implementation of that feature in code. But it may also have unintended consequences for the construction of the software. It might, for example, make it more difficult for other features to be implemented; or it might require more time, effort and money to implement than had been anticipated.

Of course, not all unintended consequences are negative. For example, it might turn out that part of the code created to implement a given design feature can be re-used to help implement other features. Often, this potential for re-use is not recognized until implementation of the given feature is well underway.

A given design feature might have unforeseen consequences for the use of the software. Such consequences might include conflict with a user goal that had not originally been included in the list of requirements. Another negative consequence for use might be unforeseen effects on user behavior. For example, it might have originally been thought valuable to give users considerable control over the visual appearance of documents they create with the developed software. But if this leads to users spending excessive amounts of time designing documents for internal distribution, then the requirements might have to be modified.

There are myriad additional reasons why certain consequences of a decision only become known after that decision is made. The crucial point is that once these consequences become known they provide additional rationale about previously made decisions. They can provide the basis for additional arguments, evaluation criteria, assessments of decision alternatives with respect to criteria, additional decision alternatives to consider or even additional decision tasks. If this additional rationale could significantly change the quality of the software artifact or the cost of its creation, the rationale may well motivate decision makers to re-open already settled decision tasks—thus producing iteration in the decision making process.

16.6.2 Principles for Rationale Approaches to Support Iterative Development

16.6.2.1 A Conceptual Framework for Iterative Reasoning

Certain approaches to rationale in SE are explicitly based on theories of iterative development. Scenario-Claims Analysis, for example, is based on a theory of "the task-artifact cycle"; and the WinWin approach to rationale is based on Boehm's Spiral Model of software development (Boehm and Kitapci 2006). But such approaches tend to be quite specialized. For example, the former deals only with useage-centric rationale and human-computer interaction, while the latter deals only with one of the many methods for iterative software development. These approaches demonstrate the possibility of integrating rationale with iterative development, but they do not provide a sufficiently general conceptual framework for doing so.

Above all, a general conceptual framework should describe what is required of a rationale approach for it to be usable in iterative development. This description should be specific enough not only to decide whether a given rationale approach is adaptable for use in iterative development, but also to indicate how it would need to be adapted.

Of special interest is what a conceptual framework has to say about the use in iterative development of the many rationale approaches that fail to indicate how they might support such development. These include IBIS, PHI, Potts-Bruns, QOC, DRL and RATSpeak. All of these approaches model rationale entirely around the concept of *planning,* in the sense of reasoning about how to act before action takes place. This model is not compatible with iterative development, in which decisions about requirements and design lead to action in the form of implementation and use, which in turn produce rationale that informs further decisions about requirements and design.

Contrasting with the model of rationale as planning is Schön's model of rationale as *reflection-in-action* (Schön 1983). The latter involves observing the consequences of actions and then reflecting on, i.e. reasoning about, how to modify future actions in view of these consequences. Schön's theory, which he calls Reflective Practice, models practical reasoning as an iterative process of learning through action. This theory, when combined with planning, provides precisely the foundation needed for a conceptual framework for rationale in iterative software development. To be more precise, if decision making starts as planning and then follows up with reflection-in-action when the less-than-adequate consequences of planned decisions are discovered, the result is a model of reasoning that fits interactive software development.

16.6.2.2 Features of Rationale that Support Iterative Development

The features of rationale that support iterative development are of two types: required and desirable. The former refers to things without which rationale simply does not support iterative development. The latter refers to things that provide richer levels of support for such development. The discussion below starts with the required features.

Decision-making is not a one-shot process. The single, most basic feature a rationale approach must have if it is to support iterative software development is that it must allow the re-opening, re-deliberation and re-deciding of previously decided decision tasks. These tasks include the determination of what the requirements for the software are, what its design features are and how these are implemented.

Feedback is not inhibited. The second requirement is that the rationale approach should not inhibit the recording of feedback, because feedback is the most important source of the rationale for iterative development. This might sound like a trivially obvious requirement until one realizes that nearly every existing approach to rationale violates it. In particular, as they are practiced, almost all decision-centric approaches mandate *de facto* a sequence in which elements are recorded. For example, they require that decision tasks be recorded before decision alternatives are recorded and that decision alternatives be recorded before evaluations are recorded, e.g. before evaluation criteria or evaluative arguments are listed. The problem with these mandated sequences is that feedback can easily take the form of a piece of rationale that is disallowed by a mandated sequence. For example, feedback might take the form of an idea for a design feature (an *option* in QOC or a *position* in IBIS) that does not respond to an already stated decision task (a *question* in QOC or an *issue* in IBIS). If feedback is not to be inhibited, *it should be possible to record elements in any sequence in which they might arise from feedback.*

The full spectrum of SER is documented. In addition to the above-listed requirements, there are possible features of rationale approaches that provide additional support for the representation and aid of iterative development. These include the documentation for SE activities other than design. The documentation of the rationale for requirements determination enables the representation of the revision of requirements. Given the crucial role that the volatility of requirements plays in iterative design (Rajlich 2006), this especially important. One crucial source of feedback, and thus of the rationale for iterative development, comes from the experience of software use. Documentation of usage-centric rationale, such as is provided by Scenario-Claims analysis, thus provides additional support for iterative design. Additional feedback comes from construction and maintenance. Documentation of the decision-centric rationale for these activities provides additional support for iterative design, because insights resulting from these activities can become the basis of rationale for re-thinking decisions about requirements, design, and construction. In short, *the more a rationale approach supports the documentation of the full spectrum of software engineering rationale (SER), the more it supports iterative software development.*

Influence/dependency relationships are documented. Another way in which a rationale approach can support iterative development is by representing *influence relationships*—and/or their converse, *dependency relationships*—between various SE tasks. These include relationships between decision tasks, especially between *different types* of decisions, including decisions about requirements, design, construction, maintenance and testing. The cycle-rich network of these relationships is the mechanism that drives iteration in development. In addition, this network is crucial for determining the impacts of changes, which are both a consequence and a cause of iterative development.

There are several distinct types of influence/dependency relationships that are of interest. One has to do with the way in which one decision influences the making of another decision. Typically, this relationship exists when the former decision helps either to generate or to evaluate decision alternatives for the latter.

A second type of relationship exists when one decision *raises*—or *leads to*—another decision task. So, for example, deciding that D is to be a design feature of a software artifact leads to the decision as to how to implement D. In this case the latter decision task *presupposes* the decision to have D as a design feature. If it is later decided that D should not be a design feature, i.e. the presupposition becomes false, the latter decision task *ceases to be relevant* to the project.

A third type of relationship exists when experiences with a task provide reasons for re-visiting previously settled decision tasks. For example, failure of users to figure out how to use the software to accomplish a required task could provide a reason for re-examining implementation decisions or design decisions. On the other hand, the implementation of a design feature in code might reveal that parts of this code could be re-used for other purposes. This might result in change of other implementation decisions. It might also suggest that additional functionality could be implemented with very little additional effort. This in turn might suggest that a decision about requirements be revisited to include some of this additional functionality.

There are many other ways in which experiences with tasks can provide reasons—and rationale—for previously settled decision tasks. In fact, there are at least as many ways as there are types of elements in whatever rationale schema is being used in the given rationale approach. So, for example, in IBIS these reasons (this feedback) might take the form of a new issue (decision task), a new position (decision alternative), or a new argument. In a rationale approach based on a more complex schema, such as DRL, there will be additional differentiation in the roles that feedback can play. DRL, for example, enables feedback to also be in the form of a goal that can serve as an evaluation criterion on a decision alternative.

The epistemological status of rationale is documented. To deal with the iterative processes central to many of the more recent approaches to SE, existing decision-centric rationale approaches need to be modified. Fortunately, the modifications needed to implement the features described above would be straightforward and relatively easy to implement. There is one crucial caveat, however. Arguments that are based on feedback from implementation and use often have a different epistemological status than the arguments made prior to implementation and use. The latter largely consist of predictions—i.e. hypotheses—about the *possible* consequences of action. The former describe the *actual* consequences of action. When there is a conflict between them we would generally expect that reports of actual events will be taken as refuting the predictions. Even if feedback does not conflict with any argument, it might introduce evaluation criteria that were not considered in the original decision. This means that if feedback argues against the decision on an issue, that issue probably needs to be re-opened.

The crucial point here is that the two different kinds of argument generally have different levels of credibility. One kind contains speculative predictions of consequences; the other reports of actual consequences. This asymmetry in credibility raises the question of whether it is misleading to represent them the same way in the rationale schema. It may be important to indicate whether arguments are predictions or tests of predictions. It might even be important to indicate the source of the feedback; and doing this would require only minor modifications in the schemas for decision-centric rationale approaches. If a rationale approach documents the full spectrum of SER and supports all the dependency relationships described above, this might by itself provide sufficient indication of the sources of rationale to determine its credibility.

16.6.3 Supporting Iterative Development by Combining Decision-centric and Usage-centric Rationale

Usage-centric rationale can be an important driver of iteration in SE. For decision-centric rationale to reach its full potential in SE, it needs to be augmented and integrated with usage-centric rationale. Doing this requires using the rationale from such methods as SCA to inform the evaluation of decisions in decision-centric approaches. In using rationale derived from actual usage, its fundamentally empirical character means that it generally has higher credibility than evaluations based on hypotheses about the consequences of decisions.

SCA's rationale about collections of decisions will only become available after those decisions are made. It will therefore take the form of feedback that either confirms those decisions or challenges them and forces them to be reconsidered.

SCA's evaluation schema strongly resembles the criterion-based evaluation schemas of QOC and DRL. All three of these methods deal with the evaluation of explicit system features based on stated criteria, with these evaluations being either positive or negative. If the features described in SCA match the decision alternatives in QOC or DRL, connecting SCA's evaluations to QOC's and DRL's as feedback should be straightforward.

16.7 Challenges to Rationale Usage

16.7.1 Solving the Capture Problem

16.7.1.1 The Capture Problem

By far the greatest challenge to making rationale usage practical is *the capture problem* (Conklin and Burgess-Yakemovic 1996), i.e. the fact that that it has proved surprisingly difficult to capture rationale in real-world projects. This is not to say that rationale capture has not been successful, but rather to point out that the conditions under which has been successful are either hard to achieve or not well understood.

The most common cases where rationale capture has worked are when there are *champions of rationale usage* within a project team or when professional rationale documenters or professional documentation facilitators are available (Conklin and Burgess-Yakemovic 1996). Unfortunately such champions tend to be in short supply, and the people who fund projects often do not see the value of paying for professional rationale documenters or facilitators. When rationale champions and professionals are not present, the documentation of rationale has typically been left to those who participate in decision making. To date, this has typically meant designers. Unfortunately, these designers have largely resisted documenting their rationale.

16.7.1.2 Analysis of the Problem

There are a number of possible explanations for resistance to rationale capture. Some researchers point to the *intrusiveness* of rationale capture as the problem. One kind of intrusiveness is due to the work required for capture. Most capture involves designers writing up their rationale in a given rationale schema. This requires a great deal of work in addition to the normal work of design.

Other reasons for resistance to capture can include political and legal factors. Designers might not want their bosses or the public to know the real reasons for their decisions. They might also want to protect themselves from potential law suits. And there is the problem that any argument for a decision can become a double-edged sword that provides others with a way to attack decisions made.

For descriptive uses of rationale, motivating rationale capture can be a fundamental problem, because, *by definition, the rationale recorded does not aid those who do the work of recording it.* In other words, descriptive approaches run afoul of Grudin's principle that collaborative systems tend to fail when those who do the work are not the beneficiaries of that work (Grudin 1988). For prescriptive approaches, Buckingham Shum and others have argued there is a benefit to decision makers from recording their rationale (Buckingham Shum et al. 2006), so they should be more motivated to do it. Yet even here rationale capture has been difficult to achieve.

Another possible reason for the failure of capture in both descriptive and prescriptive approaches is that capture might actually be detrimental to design in ways that go beyond its cost in resources. For example, Fischer et al. (1996) use Schön's theory of Reflective Practice to argue that rationale capture disrupts the intuitive aspects of designers' thinking. A more radical position is taken by Shipman and Marshall (1999b) who argue that semi-formal schemas, such as those used in most rationale approaches, are themselves the problem. As they see it, all such schemas are obstacles to information capture.

16.7.1.3 Approaches to Solving the Problem

One possible way of getting capture to work is to convince those who fund software projects of the value of rationale usage. This might merely require doing a better job of explaining or demonstrating the benefits to them. But it may require more, such as decreasing the resistance of decision makers to rationale capture, increasing the benefits of such capture—or both.

One approach to reducing resistance to rationale capture is to reduce its intrusiveness into decision making processes, either by reducing the amount of work it requires or by reducing its disruptiveness. The traditional capture process has combined capture with the formalization of rationale using a schema. A crucial insight motivating many efforts at reducing intrusiveness is that it is actually the *formalization* that takes so much time and effort. If rationale were first captured in "raw" form, it could be formalized later. This would not in itself reduce the task of formalization, but it would decompose the problem into two smaller problems. It would also enable more rationale to get recorded. Of course, raw rationale would be difficult to retrieve if not structured and indexed.

A number of strategies have been devised for capturing raw rationale in informal, i.e. schema-free forms and then using various "tricks" for reducing the effort of formalizing it. For example, Shipman and his collaborators from Xerox PARC built "spatial hypertext systems" (Shipman, Marshall 1999a) that enable informal input of information in a 2D space and then use automated routines to infer the structure of that information from its spatial arrangement. Reeves created a system that uses a schema-free approach to capture (Reeves and Shipman 1992). With his system, designers write their rationale as textual notes in the graphical representation of a model of the artifact being developed. The design history of the artifact then becomes the means by which rationale is structured. A different schema-free and completely non-intrusive approach is used by Myers, Zumel, and Garcia (1999) (See Chapter 4 of this book, Learning from Rationale Research in Other Domains). They add semantic information to a CAD system's symbol library and then infer the design rationale from the designer's use of the system. This approach, however, does not produce argumentation as such. Another schema-free approach is to capture the rationale that is naturally elicited as part of informal project communication. In this case, eliciting rationale is not an extra task for decision makers. It is instead a normal and accepted part of the process of collaboration. Completely automated approaches might then be used to structure this rationale, for example, by using natural language processing (McCall and Mistrik 2005). Alternatively, semi-automated approaches can be used such as Shipman's *incremental formalization* (Shipman and McCall 1994).

One approach to reducing the cognitive overhead of capture is to use the strategy of *differential description*, in which designers only need to describe how the rationale for the current project differs from other rationale. One way to do this uses domain-oriented issue bases in PHI (Fischer et al. 1996). These contain rationale commonly used in projects in a given domain, including commonly raised issues, positions and arguments. Decision makers then only need to add the information missing, including their decisions on the issues.

There are other ways in which differential description might be implemented. One would be by using rationale-annotated cases of similar projects, such as those provided by the ARCHIE system (Zimring et al. 1995) (See Chapter 4, Learning from Rationale Research in Other Domains.) Another way might be to use design patterns annotated with rationale (Pena-Mora and Vadhavkar 1996). Of course, differential description only works for domains where previous design work has been done and where someone has built collections of issue-based discussion, precedent cases or design patterns. By definition, this approach is not useful for unprecedented problems.

16.7.2 Solving the Delivery Problem

To date, almost all delivery of rationale to those who need it has been done using hypertext-based information systems. One problem with this approach is that potential users of such a system generally do not search for information unless they think that there is information in it worth searching for. But how are they to know that such information exists if they do not search for it? If new information that would be useful for a given user is input into such a system, how does that user find out about this?

Hypertext systems have a partial answer to such questions in the form of *associative indexing*, i.e. indexing by linking to other information. This enables new information to be discovered by being linked to other information that a user knows is relevant to their current concerns. Thus, for example, a link might help a user discover a newly created argument against a decision alternative that they favor.

The potential difficulty of the link-based approach is that users do not discover the link if they are not already using the system. To assure the discovery of new and relevant information requires not only the users be using the system but also looking at the information to which the new information is linked.

One partial solution to this problem is to integrate the rationale management system into the software used for SE. This would enable the hybrid system to alert software engineers to existence of links to rationale that is relevant to the SE tasks that they undertake when those tasks are at hand. This is the approach that Burge has used in linking rationale to source code being edited in the Eclipse IDE (Burge and Brown 2006). As programmers browse through the code, they are alerted to the existence of links to rationale relevant to the sections of code they examine.

Additional functionality may well be needed to compensate for the limitations of hypertext systems. One example of such functionality is to provide knowledge-based agents that can alert users to the existence of rationale relevant to their concerns. Fischer et al. have used this approach (Fischer et al. 1996) as have McCall and Johnson (1997), but much more research in this area is needed. Making this approach successful may require research on modeling stakeholders in SE to understand their concerns and what rationale is relevant to these concerns.

16.8 Summary and Conclusions

The conceptual framework presented in this chapter has attempted to describe the concepts and ideas that connect rationale to SE. The intention has been to show both the potential of rationale to serve the goals of SE and the challenges to successful use of rationale in software projects. The framework has described the roles of both decision-centric and usage-centric rationale approaches in SE. Included in this discussion were descriptions of limitations and advantages of rationale approaches for different aspects of the SE process. The framework has also described the modifications to decision-centric rationale that are needed to make them fully serve the goals of the iterative approaches to SE that have gained popularity in recent years. Finally, it has identified and explained the two crucial challenges to successful rationale usage in practical projects: the capture problem and the delivery problem. The purpose of the conceptual framework is to provide a guide for practical use of rationale in real-world software projects and for research on how to improve rationale applications.

17 An Architectural Framework

A rationale-based approach to software engineering requires rationale management systems that can integrate the many types of rationale with each other and with the processes of creating software engineering artifacts. Accomplishing this integration in turn requires that such systems be actively connected with software engineering tools, external communication sources and persistent stores of reusable rationale. This chapter describes an architectural framework for such integrative rationale management systems.

17.1 Introduction

17.1.1 An Integrative Architecture for Rationale-Based Software Engineering

Fully implementing rationale-based software engineering (RBSE) will require the creation of software that can manage rationale effectively to support software engineering (SE) practice. Such a rationale management system (RMS) needs to be able to elicit and to record large amounts of useful rationale, to structure it for ease of comprehension, to index it for retrieval and to deliver it to those who need when they need it. Furthermore, the system needs to do all this in ways that are compatible with SE processes. This chapter analyzes what this implies for the architecture of RMSs that support RBSE. The result of this analysis is a set of recommendations in the form of an architectural framework for RBSE.

An RMS must be able to do three things if it is to support RBSE. The first is that it must represent all the various types of rationale that occur in software projects in a form that supports SE. The other two are that it must make substantial progress in alleviating the rationale capture and rationale delivery problems. Above all, the capture problem must be effectively dealt with, for without the ability to capture adequate amounts and types of software engineering rationale (SER) there will be little value in representing and delivering rationale.

The basic approach recommended here for achieving the above-stated goals and requirements is to use *an integrative architecture* for RMSs. This architecture is integrative in two respects. One is that it integrates the many different types of SER with each other. The other is that it integrates the processes of capturing, structuring and delivering rationale with the processes of SE, which are largely centered on the creation of various SE artifacts, including documents, models and code. The former type of integration weaves the myriad types of SER into a single argumentative structure that produces the final software artifacts. The latter type of integration improves the quantity and quality of the rationale that is captured and delivered.

This chapter will not attempt to describe a complete software architecture for an RMS that supports RBSE. Instead it will describe an architectural framework for such systems. This framework consists of an abstract description of the essential, common characteristics of an integrative architecture for RMSs, leaving the "accidental" specifics of the architectural design to others. While the term *architectural framework* is often used in the object-oriented sense as referring to a specific set of classes, the term is used here in a looser sense to mean a more informal description of the main features of a software architecture.

17.1.2 Objectives of this Chapter

The main objectives of this chapter are 1) to describe an architectural framework for RMSs capable of implementing a rationale-based approach to SE and 2) to explain the reasons for its design. Section 17.2 explains the need for an integrative approach to rationale management to represent and to integrate all the various types of SER and to alleviate the problems of rationale capture and delivery. Section 17.3 describes the integrative, architectural framework itself, starting with an overview of the framework in Section 17.3.1. It then describes the workings of the RMS system in Section 17.3.2 and its connections of to external systems and sources in Sections 17.3.3. Finally, Section 17.4 summarizes the chapter and draws conclusions about the use and significance RMSs built using the architectural framework.

17.2 The Need for an Integrative Approach to Rationale Management

17.2.1 Representing and Integrating All Types of Software Engineering Rationale

RBSE, by definition, involves the use of the full spectrum of software engineering rationale (SER). To support RBSE, an RMS must therefore be capable of simultaneously modeling the rationale for every activity of SE, including the activities of requirements determination, design, construction, testing, maintenance, project management and even the use of the software. But this is not enough. It must also be capable of modeling the various relationships that integrate these different types of rationale into a single network of reasoning that results in the code given to users. This involves not only integrating various types of decision-centric rationale, but also integrating these with usage-centric rationale.

17.2.2 Alleviating the Capture and Delivery Problems

17.2.2.1 The Disconnect between Rationale Management and Software Engineering

The goal of SER research is to use rationale approaches to aid SE. To date, however, there is has been only sporadic and modest success in achieving this goal. While positive results have been reported in some notable cases, e.g. (Conklin and Burgess-Yakemovic 1996; Buckingham Shum et al. 2006), it is widely believed among researchers that the effort to achieve this goal has run into fundamental difficulties, especially in the form of the rationale capture and delivery problems. Any effort to create RMSs that can alleviate these difficulties needs be based on ideas about their causes and how to overcome them.

The position taken in this chapter is that rationale approaches and management systems generally have not done enough to fit into and support the practices that software engineers use in developing and maintaining software systems. Currently, decision making in software projects is accomplished through the use of various SE tools combined with informal communication among project participants. Rationale approaches and RMSs have often been presented as alternative means for decision making, with RMSs being used instead of SE tools and informal communication being replaced with communication structured according to a rationale schema. This chapter explores a different strategy in which rationale approaches and systems are used to support rather than supplant existing approaches to decision making in SE. It is argued here that this strategy offers the potential of substantially alleviating both the rationale capture and rationale delivery problems.

Any attempt to support SE practice must be based on an awareness of its artifact-centered nature. Almost all SE processes, methods and tools are aimed at the production of special types of SE artifacts. One type of such artifacts is executable code, such as prototypes and various versions of the software product being created. But there are also many non-executable artifacts that are used as means for devising code, including documents and models of various types. Potts and Bruns (1988) first described the crucial role of such *intermediate artifacts* in software design. Among these artifacts they list, "informal documents describing the functional specification of the system, architectural sketches, detailed designs, pseudo-code, structure diagrams, or formal specifications" (Potts and Bruns 1988). By broadening the scope of rationale from design to the entire spectrum of SE activities, as this book does, the number and variety of such artifacts are increased substantially.

Decision tasks in SE generally arise out of the desire to create SE artifacts. For rationale approaches and systems to support SE practices, they must contribute to the handling of such decision tasks by engaging with and supporting the use of SE tools and discussion among project participants. In particular, RMSs should 1) capture rationale from discussion and SE tool use and 2) deliver rationale that informs discussion and the use of SE tools to make decisions.

Unfortunately, many of the RMSs that have been proposed for use in SE are monolithic, stand-alone systems and, as such, have no computational interaction or connection with the SE tools or project discussion that are used to create SE artifacts. Such RMSs are literally *out of the loop*, and thus never come into play in the processes of creating SE artifacts.

The irony here is that most rationale approaches and SE processes have a strong conceptual connection in their common focus on decision making processes. But there is no way to exploit this conceptual connection when RMSs do not have a tangible, computational connection to the creation of SE artifacts. Without this connection there is no way to capture rationale during SE decision making, which is when it is generated; and there is no way to deliver rationale during this decision making, which it is when it is needed.

17.2.2.2 Integrating Rationale Management with Software Engineering Decision Making

For any RMS to be successful in capturing and delivering the rationale for SE decision making, it must be integrated into the artifact-centered decision making in SE. The only way to guarantee that this happens is to represent the decision making about SE artifact in the rationale. Among the domain-independent approaches to rationale, such as IBIS, PHI, QOC and DRL, the only time decisions about artifacts are made is in the case of design-space decisions; and in fact, QOC is the only rationale approach that guarantees that such decisions are dealt with. For rationale to be integrated with the process of making decisions about SE artifacts those processes must be represented in the rationale—as what might be called *SE artifact-space analysis* by analogy with QOC's *design-space analysis*. The greater the number of artifact-space decisions represented, the more rationale process are integrated with SE processes. The integration is complete if the set of artifact-space decisions describe all the SE artifacts. This is, in fact, very close to what Potts and Bruns proposed when they advocated the incorporation of representations of SE artifacts into rationale hyperdocuments (Potts and Bruns 1988; Potts 1996).

It is, however, not enough for an RMS merely to represent decisions about artifacts. The RMS should be able to guarantee that the state of the representation of decisions in the rationale always matches the decisions about artifacts made with SE tools. But this can only happen if there is some sort of computational connection between the RMS and the tools that guarantees 1) that a decision made with a tool is immediately updated in the rationale and 2) that an artifact-space decision made in the RMS is immediately updated in the SE tool. Even this is not enough. Any new decision task undertaken using an SE tool must immediately get represented in the RMS, and vice versa. In fact, to whatever extent the elements and relationships of the rationale schema are explicitly dealt with by SE tools, there must be the same sort of mutual updating so the representation of the decision making processes in the rationale matches the state of these processes in the SE tools.

17.2.2.3 An Integrative Approach to Capturing Rationale

An integrative approach to rationale capture is one that enables capture during the creation of SE artifacts. To the extent that such artifacts are produced using SE software tools, this means that it must be possible to capture rationale about a decision task while that task is being accomplished using the tool. To the extent that these artifacts are created by means of communication among project participants, an effort should be made to capture this communication, because it provides a valuable source of project rationale. Such capture can be accomplished by making recordings or written records of meetings and computer-mediated communication.

Though extracting and structuring rationale from records of communication presents challenges, it also has decisive advantages for alleviating the capture problem. The reason is that, unlike almost all other modes of capture, stating rationale in communications to other project participants is not perceived by decision makers as extra work beyond the normal work of decision making. This is because such communication is the central means by which collaboration in groups takes place. The consequence is that decision makers tend not to resist stating rationale as part of collaborative communication.

Three major modes of rationale capture should be possible: unprompted, prompted and automated. *Unprompted capture* means that the person stating the rationale spontaneously decides to enter rationale of some type. *Prompted capture* means that the person using the tool states rationale is response to a prompt of some type, e.g. in response to a statement by someone else or a request for rationale of a certain type. Both informal, unstructured rationale input and schema-based rationale input should be supported in both unprompted and prompted capture.

To the extent that the processes of making decisions about SE artifacts are represented in the rationale and automatically updated by the RMS in the manner described in Section 17.2.2.2, *the rationale for decision making in SE will be automatically captured by the RMS*. Decisions and decision tasks are likely to be captured in this way. Dependency relationships might also be captured automatically. But verbal argumentation is likely to be captured only by the decision makers voluntarily entering this argumentation or by mining records of communications between project personnel for relevant argumentation. Though the capture of all the relevant rationale is generally not possible using automated techniques, the amount of rationale that can be captured in this way should greatly reduce the burden on decision makers for documenting their rationale.

17.2.2.4 An Integrative Approach to Structuring Rationale

An integrative approach to structuring rationale is one that enables structuring to take place during the creation of SE artifacts. While both unprompted and prompted modes of structuring should be supported, the main opportunities for reducing the work of structuring come from the use of automated means. To the extent that decision making is tool-based, there are opportunities for structuring rationale by associating it with decisions and artifacts, i.e. by *decision-based indexing* and *artifact-based indexing*. The former can be accomplished by automatically linking rationale to automatically generated representations of the decisions being made; the latter by linking rationale to the particular artifact being created or modified. In addition, by keeping a version history of the decisions, rationale can be further structured by associating it with a particular moment in that history. Dependency relationships among decisions and among artifacts might also be captured automatically and used to structure the rationale associated with those decisions and artifacts.

To the extent that rationale is part of project communication, the inherent structure of that communication can be used to automatically structure rationale. For example, communication typically involves turn-taking, and this can be used to give a basic structure to rationale. Threaded discussion provides additional structure. Structuring within individual textual "utterances" can to some extent be done using natural language processing techniques, e.g. as in the work of McCall and Mistrik (2005), though this research is still in its early stages. Finally, structuring in the form of linking to relevant keywords and subject headings can be done using well-establish techniques of information retrieval.

17.2.2.5 An Integrative Approach to Delivering Rationale

An integrative approach to delivering rationale is one that enables delivery to take place during the creation of SE artifacts. To the extent that SE artifacts are created using SE tools this implies not only delivery of rationale during the use of tools but also delivery of rationale relevant to the use of those tools in the decision making about SE artifacts. Decision-based indexing and artifact-based indexing play a decisive role in enabling this integrative approach to delivery. To the extent that project communication is computer-mediated, delivery should be possible by means of the communication systems being used. Providing rationale relevant to the rationale contained in communications between project participants requires the ability to understand the content of that rationale. It has not yet been adequately demonstrated how to do this automatically; so this aspect of integrative delivery will have to wait for such a demonstration.

The way in which integrative delivery of rationale can help to solve the delivery problem is by going beyond the traditional approach of browse-and-query. This approach requires the person doing the browsing and query-based searching to know that they need information (rationale), that the needed information is in the documented rationale and how to retrieve that information. Unfortunately, it is common not to know these things.

The way integrative delivery can help is by using the nature of the decision task at hand and the identity of the artifact being created to do two things:

1. alert project personnel to the availability of documented rationale that they have not yet seen but that is relevant to the decision task at hand or to the artifact they are currently creating
2. retrieve and display that rationale.

A crucial point about the delivery of relevant rationale is that there is no reason to restrict where this rationale comes from. In addition to looking for useful rationale in the documented rationale for the current project, it might well be that such rationale can be retrieved from other sources. In particular, there are a number of approaches to creating persistent stores of reusable rationale, including pattern-based approaches, issues-based approaches and case-based approaches. Utilizing such external sources of rationale not only has the potential of enhancing the value of rationale delivery, it also has the potential to reduce the amount of rationale that needs to be captured and structured in the current project. The principle here is *differential description*: it is only necessary to capture the differences between the current project's rationale and the rationale retrievable from external stores. Where the current project uses rationale from external stores, the only things needed are links to that external rationale.

17.3 Framework of an Integrative Architecture for Rationale Management in Software Engineering

17.3.1 An Overview of the Framework

The architectural framework consists of a hypermedia-based RMS with connections to three types of external entities: 1) SE tools, 2) communication systems and sources, and 3) persistent stores of reusable rationale. The RMS itself manages hyperdocuments containing linked collections of rationale nodes and nodes representing SE artifacts. The external connections enable 1) acquisition of rationale from external systems and sources, 2) automated structuring of rationale using connections to external systems and artifacts, and 3) the delivery of rationale through external systems. The activities of the RMS and its connections are explained in Sections 17.3.1, 17.3.2, and 17.3.3.

The architectural framework is integrative not in the sense of requiring integration but rather in the sense of facilitating it. There are two types of integration that the architectural framework facilitates. One is the integration of the rationale for all the various activities of SE, including requirements determination, design, construction, testing, use and maintenance. The other is the integration of the creation and use of rationale with the creation and use of SE artifacts.

The architectural framework dictates that the RMS be capable of managing a linked collection of hyperdocuments associated with different SE activities. For each such activity, the framework enables the construction of the sort of hybrid hyperdocuments of both rationale and artifact nodes first suggested by Potts and Bruns (1988) and later elaborated by Potts (1996).

But the hybrid hyperdocuments proposed here go beyond those of Potts and Bruns in two important respects. One is that, in addition to representing the sorts of intermediate artifacts that Potts and Bruns discussed, the new hybrids can also represent executable artifacts, i.e. code. The second respect in which the hybrids proposed here are different is that they enable the computational coupling of hyperdocument nodes and links to the parts and structure of actual SE artifacts. This coupling makes possible the automatic capture and structuring of rationale 1) through the use of SE tools and 2) from records of communication between project participants. It also makes possible the delivery of relevant rationale during the use of tools and computer-mediated communication.

17.3.2 Workings of the Rational Management System

17.3.2.1 Representation

To fully support RBSE, the RMS must represent the rationale and associated artifacts for every aspect of SE, including requirements engineering, design, construction, testing, use maintenance and project management. Since the set of artifacts associated with each aspect of SE is likely to be different, and since different aspects of SE might use different SE methods, it must be possible to use a different schema in representing the rationale and artifacts for each aspect. To integrate the various aspects of SE into a coherent overall SE process, it must also be possible to establish links between the models of rationale for the individual aspects. In particular, it must be possible to establish dependency relationships between the various aspects and to support these relationships with computation.

The RMS should be capable of constructing hyperdocuments for all of the schema-based, argumentative rationale approaches currently found in the literature. These include IBIS (Kunz and Rittel 1970), PHI (McCall 1990), QOC (MacLean et al. 1991), DRL (Lee 1991), RATSpeak (Burge and Brown 2006) and Scenario-Claims Analysis (Carroll and Rosson 1996), as well as the various SE-specific approaches, such as WinWin (Boehm and Kitapci 2006), TEAM (Lacaze et al. 2006), REMAP (Ramesh and Dhar 1992). In addition, the RMS must make it possible for software engineers to invent new schemas and to arbitrarily modify schemas to accommodate information that is specific to particular software projects, SE aspects, SE artifacts, SE methods and the problem-solving styles of software engineers.

The RMS should have the ability to create typed and labeled links and nodes with content in every major type of medium, including text, sound, 2D and 3D graphics and animation. It must be possible to establish links not only between nodes but also between nodes and links. One specific reason for doing this is in order to be able to represent rationale approaches like QOC that require this. The more general reason is because, as Lee has pointed out (Lee 1991), links correspond to claims. Since they are claims, it should be possible to comment on them and reason about them in various ways. This requires linking rationale nodes to the links being discussed.

The RMS should be an open hypermedia system with the capability of associating nodes and links with external content created in external systems. In particular, it should be possible to use external content as the content of nodes. It also should be possible to link directly to eternal content.

17.3.2.2 Computation

The central mechanism for realizing the integration that is the hallmark of the integrative architecture is the use of dependency relationships, including both ordinary links and computed dependencies. These relationships and their computational support integrate the collections of rationale for different SER activities with each other and with the artifact-centered decision making processes of SE.

To support integration, the RMS needs have the capability of establishing and supporting computable dependencies between the states of different nodes. It should be possible to use any algorithm to compute these dependencies. Ideally, there should support for users establishing and editing basic computable dependencies, such as those based on algebraic formulas and conditional statements.

Supporting integration also requires that the RMS should support traceability of both computed and non-computed dependency relationships. The RMS should also support what-if computation with computable dependencies. It must also be possible to establish computable dependencies of internal content on external content. Where external systems allow it, it should also be possible to establish computational dependencies of external content on content of RMS hyperdocuments.

17.3.2.3 Display and Input

To support the use of various rationale approaches, the RMS should provide standard hypermedia display capabilities, including outline-formatted display of node structure and content in the manner of JANUS (Fischer et al. 1996) and PHIDIAS (McCall et al. 1994) as well as graph-based displays in the manner of gIBIS (Conklin and Begeman 1988), SIBYL (Lee 1990) and Compendium (Shum et al. 2006). The RMS should be able to alert users to existence of rationale associated with a particular decision task, artifact or condition and then display the relevant rationale.

As is typical of hypermedia systems, the input of content to hyperdocuments should be possible using editors for various media, these editors being part of the RMS or external systems. Editors should be provided for node content, hyperdocument structure and schemas. To support all major approaches to capture, both prompted and unprompted input should be possible. Both schema-driven and free-form structuring of input should also be supported.

17.3.2.4 Additional Capabilities

To support SE practice, the RMS should enable multi-user creation, editing and display of representations of rationale and artifacts as well as hyperdocument structure and schemas. To do this, it must provide communication and shared workspaces for members of groups of project participants—e.g. members of a development team working on the design of a particular subsystem.

To support an integrative approach to capture, structuring and delivery of rationale, the RMS must also support the creation and browsing of a version history of the creation of hyperdocuments. It should be possible to attach rationale and commentary at any point in the version history.

17.3.3 Integration with External Systems

17.3.3.1 Integration with Software Engineering Tools

To support an integrative approach to capture, structuring and delivery of rationale, one type of connection that should be possible between a hyperdocument and an artifact is a computationally coupling of nodes with SE artifacts, including those artifacts that are parts of other artifacts. This coupling means that a change in the state of an SE artifact, such as a part of document or model, can automatically result in a change in the state of a node in the hyperdocument. This coupling enables the content and existence of the actual artifact to be reflected automatically in the content and existence of a node that represents it. In other words, the creation, deletion and change of content of an artifact, or part of an artifact, could automatically be reflected in the creation, deletion or change in content of corresponding node. Coupling in the other direction would mean that a change in the node is reflected in a change in the corresponding artifact. This coupling is likely to be harder to achieve but is useful where possible.

The second type of connection that should be possible between a hyperdocument and an SE artifact is that the structure of the artifact should be coupled with the links in such a way that a change in the structure of the artifact is automatically reflected in the structural connections between nodes in the hyperdocument. Coupling in the other direction is useful but likely to be harder to achieve.

The computational coupling of hyperdocuments to artifacts is the mechanism that enables the automatic capture and structuring of rationale from the use of software tools in making SE decisions. In particular, it enables any decision tasks, decision alternatives and final decisions to be reflected automatically in corresponding hyperdocument nodes with appropriate links between these nodes. It also enables the automatic modeling of the state of the artifacts at any given time. When combined with the version history capability of the RMS, this makes it possible to have a history of the evolution of the SE artifact as it is created and modified. Such a history by itself suggests much of the rationale for the final form of the artifact, but is also provides a useful way of automatically structuring and indexing rationale by the states of the artifact's evolution.

17.3.3.2 Integration with Communications Systems and Sources

While the RMS needs to provide communication capabilities for group creation of rationale through argumentative discourse, these capabilities cannot fully satisfy the communication needs for a project group. In particular, it is naïve to suppose that all group communication can be mediated by structured argumentative discourse. There need to be multiple additional channels for communication, including informal discussion and meetings. Some of this communication is likely to be computer mediated, if for no other reason than that an increasing percentage of all human communication is computer mediated. To the extent that project-related communication is computer mediated, it is a near certainly that it will involve discussion that includes a substantial amount of project-related rationale. Given the current difficulty of capturing rationale, this communication is a valuable source of rationale, although mining records of communication for this rationale presents a number of technical challenges. The most accessible form of computer-mediated communication is text and may involve email, chat or other modes of text-based communication. Audio and video-based communication is more difficult to access, but still of potentially great value as a record of project decisions and the reasoning underlying them.

Some important communication is face-to-face rather than computer mediated. Meetings are the most important example. But even here, digital records of this communication are easy to make in the form of text, audio or video. Such records may well constitute important records of the history and rationale of a software project. Currently, audio and video records of face-to-face communication need to be analyzed manually; text, however, can be partially analyzed using automated or computer-assisted means. In the future, of course, analysis of audio and video will also be more computer-supported. While indexing and structuring such records is difficult and possibly labor intensive, there is no doubt that these records contain large amounts of project rationale.

To support integration of rationale management with the processes of SE decision-making, the RMS should incorporate automated or semi-automated techniques for mining records of communications among project personnel for relevant rationale. This support should include means for analyzing and indexing records of meetings and computer-mediated communication. Since the techniques for this sort of mining of communication are still in their infancy, the further description of the required functionality remains a task for future research.

Ideally support for integration of rationale management with SE decision making should also provide support for the delivery of rationale that is relevant to computer-mediated communication. But once again, this is a task for future research.

17.3.3.3 Integration with Persistent Stores of Reusable Rationale

The third major type of connection between the RMS and external systems is the linkage to external stores of re-usable rationale. There are two major reasons for this linkage. The first is to improve SE decision making by informing it with rationale that project participants would not think of on their own. The second reason is that retrieval of relevant rationale from external stores offers the potential of alleviating the rationale capture problem by obviating the need for the capture of some rationale.

The main functionality needed for retrieving rationale from external stores is the ability to browse and to query the systems which manage external stores of rationale. The sorts of queries that are useful are those that can retrieve rationale that is relevant to decision tasks the current software project is attempting to deal with. The RMS should also provide means for its users to select and record which search results are relevant to the current project and to link such rationale to the rationale for this project.

17.4 Summary and Conclusions

Implementing a rationale-based approach to software engineering requires the use of rationale management systems having an integrative architecture. Such an architecture makes two types of integration possible. One type is the integration with each other of the rationale associated with different software engineering activities, including requirements determination, design, construction, use, maintenance and project management. This integration uses dependency relationships to organize the different collections rationale into an integral body of reasoning that shapes the code that is delivered to customers. The other type is the integration of the processes of creating, structuring and delivering rationale with the processes of creating software engineering artifacts, including documents, models and code. This type of integration makes it easier to capture, structure and deliver large quantities of software engineering rationale.

The architectural framework presented here basically describes a conventional hypermedia-based RMS with a few added capabilities. It is these added capabilities that are responsible for the integration that is the hallmark of the architecture. The two crucial capabilities are 1) the ability to connect to external systems and sources of information and 2) the provision of computational support for static and computed dependency relationships. While these capabilities may sound simple, they might not be simple to implement. Nevertheless, their implementation is likely to be crucial for the success of a rationale-based approach to software engineering.

18 Rationale-Based Software Engineering: Summary and Prospect

This chapter summarizes the main points of this book and looks at the prospects for rationale to aid software engineers in dealing with the problems of future software development. It concludes that while the potential of rationale to aid software engineering is great, several crucial issues must be resolved if this potential is to be realized.

18.1 Introduction

18.1.1. Rationale as an Aid to Software Engineering

Moore's Law and the Internet have fueled an exponential explosion of technology that is unprecedented in human history. Public demand for digital technologies currently appears insatiable. As a consequence, computing and digital communication are spreading to nearly every aspect of life and to nearly every part of the world. But this technological revolution is dependent in every part and at every stage on the creation of software capable of harnessing the power of digital hardware to meet human needs. And this is where the revolution is running into trouble.

Software developers have not been able to keep up with increases in hardware capabilities, and the current rate of success for software projects is disturbingly low. Yet the demands placed on developers continue to increase relentlessly. Software is growing in scale, complexity, variety and longevity. Change in technologies and user needs is unceasing. As a consequence, software developers urgently need new approaches and tools for handling the challenges of future software projects. Rationale-based software engineering (RBSE) can play a crucial role in helping to meet these challenges.

18.1.2 Objectives of this Chapter

This book makes a case for RBSE as a crucial part of research in software engineering (SE) and as an essential part of future software development and maintenance. In previous chapters, the book has explained what RBSE is, what its potential value is for SE, what its research challenges are and how these challenges might be met. The intention of this final chapter is to provide a summary of the previous chapters and a look at the future prospects of RBSE as a way of meeting the challenges of future SE practice.

Section 18.2 presents a summary of the book that describes its overall goals and how it attempts to achieve those goals. Section 18.3 reviews some of the challenges facing future software development. Section 18.4 then looks at the potential contribution of rationale-based software engineering to meeting these challenges. Section 18.5 describes two challenges that in turn need to be met if this potential of is to be realized. Finally Section 18.6 briefly summarizes the chapter.

18.2 Summary of the Book

This book makes a case for a rationale-based approach to SE, i.e. an approach that attempts to capture and use rationale to increase the quality of SE. To do this, it explains what RBSE is, describes a wide range of ways of using rationale to aid SE, and presents frameworks meant to guide future work in the field. It also argues that RBSE provides software engineers with an invaluable tool for dealing with the increasingly difficult problems of developing and maintaining software.

Part 1 of the book introduces the basic concepts and ideas underlying RBSE. Most of the rest of the book describes issues associated with various uses of rationale in SE. Part 2 of the book describes uses for rationale in relations to such general activities as presentation, evaluation, collaboration and decision making. Part 3 describes uses of rationale in various activities within the software lifecycle: requirements engineering, design, testing, maintenance and re-use. Finally, Part 4 presents ideas meant to serve as guides for future work on RBSE, including a conceptual framework and suggestions for the architecture of rationale management systems.

The authors paint a portrait of a field of research that is just hitting its stride. It is a field that has gotten beyond the naïve mistakes of its formative years and is now appears to be converging on an understanding of its problems and how to solve them. The variety, breadth and depth of the research are considerable, and new ideas continue to emerge regularly.

The book goes to considerable lengths to survey the literature on rationale in SE and relevant other domains. But it also provides a number of new ideas. Above all, it proposes shifting the focus of research in SE from *design rationale (DR)* to *software engineering rationale (SER)*, so as to emphasize capture and use of rationale in every aspect of SE and every part of the software lifecycle. It also describes how these various types of SER might work together in the context of the overall SE process.

While the literature on SER is rich, it suffers from a sort of Tower-of-Babel of conflicting terminology. This situation makes it extremely difficult to compare the many approaches that have been proposed and applied. This book has therefore sought to devise a consistent naming scheme for the common elements and relationships of rationale without favoring any one rationale approach over the others. Basic terminology is established early in the book and then used and elaborated to create a consistent conceptual framework for discussing a variety of phenomena described in the research literature.

While staying relatively neutral, or ecumenical, in the choice of a conceptual framework for the field, this book has not maintained neutrality in all areas. Comparisons and analyses of different rationale approaches have sometimes pointed out their potential limitations or advantages. Such judgments might be controversial, but wherever the book has made them there has been attempt to provide convincing rationale for them. At very least, this rationale should provide those who disagree with those judgments a basis for arguing against them.

In the debate over the status of the rationale capture problem, the book has favored those who believe that the traditional approaches to rationale capture are not sufficient and that additional approaches are needed. In particular, in describing an architectural framework for rationale management systems (RMSs), the book has argued against the use of traditional stand-alone RMSs and in favor of systems that derive rationale from connections with SE tools, communications among project participants and external stores of reusable rationale. A similar argument is made for dealing with the problem of delivering rationale to those who need it.

18.3 The Challenges of Future Software Development

18.3.1 Managing Change

There are a number of major problems that software engineers need to solve if software development is to be successful in coping with the emerging challenges. Perhaps the most pressing of these problems is *coping with change*. The hallmark of future software creation will be change, and software engineering will itself need to change if it is to succeed.

There are two central sources of change. One is the extraordinary, continuing change in hardware capabilities. This is partly due to the explosive growth of the computational power and memory capacity of hardware due to Moore's Law, but is also taking the form of a fundamental change to parallel processing. Adding to this is the continuing growth and evolution of the Internet, which has created the possibility of a wide variety of new types of software applications.

A second major source of change is the volatility of user requirements (Rajlich 2006), though ultimately it may be the growth in technology that causes much of this volatility. The understanding of requirements can change within the time frame for developing a single version of a software product. But as products increasingly go through version after version, the change in user requirements become a major engine of the redesign of systems. Already, most of the design currently done by developers is re-design, and the need for re-design is likely to increase dramatically in coming years. Ultimately, the ongoing changes in requirements may be propelled by the fact that the satisfaction of the requirements of users and organizations fundamentally changes the environment in which they work, and this changed environment creates new needs and suggests new possibilities that lead to new requirements. Where and when this process ends—and where it is taking society in general and SE in particular—are anyone's guess.

18.3.2 Managing the Increasing Scale, Complexity and Longevity of Software Projects

Technological possibilities and customer demand are driving developers to create software of increasingly complex and diverse functionality. This in turn is leading to larger development teams with increasingly diverse types of expertise. This creates problems of coordination, collaboration and management. When project teams are small, as they have been in many well-known projects of the past, little or no formal management and communication are needed, because there is a great deal of shared, tacit knowledge. Collaboration and coordination are easily accomplished using informal communication. Management can be highly informal. In large and diverse development teams, however, there is little tacit knowledge that is shared by all team members. Coordination and collaboration are crucial but difficult. The management of such teams requires more explicit and formalized communication and procedures.

If team members do not understand how their decisions depend on decisions made by others—and vice versa—the stage is set for creation of serious errors in design, redesign, testing, implementation and maintenance. The rationale for every activity in the software lifecycle depends on other lifecycle activities. Good design depends on decisions about requirements, which may in turn depend on experiences of users of the system. Good design may also depend on experience in implementing and maintaining previous versions of a system. Similarly, decisions about maintenance and redesign require an understanding of the decisions about requirements and the previous design of the system—so that crucial functionality does not become broken as a side-effect of maintenance or redesign.

18.4 The Promise of Rationale-Based Software Engineering

The goal of rationale-based software engineering is to use rationale to improve every activity of software development and use. There are two ways it can do this: by informing these activities and by improving the reasoning processes underlying them, i.e. by making these processes more thorough, consistent and correct. Every stakeholder in a software project, including developers, clients and users, should be a potential source of rationale as well as a potential user of rationale information and methods. Every decision maker in a development team should be aware of the way in which the decisions of others in the team affect their work, especially the way such decisions have consequences for their own decision making. In addition, all decision makers should have the chance to learn from the rationale of those who have faced similar decisions in past projects. Decision makers in every activity in development, from requirements engineering to design, to testing, to implementation, to maintenance would be improved by rationale information and methods.

18.4.1 Rationale and the Management of Change

There are several ways in which rationale can help in managing change. One is by showing how decisions throughout the spectrum of SE activities depend on assumptions, requirements, and other decisions. This makes it possible to understand the both the direct and indirect effects of any changes in those assumptions, requirements or decisions. This in turn provides crucial information for deciding how to make changes and even whether they are worth making.

A second way in which rationale aids the management of change is by provide records of the intent behind the decisions that shaped the previous state of the software. This rationale helps in preserving the intent of those earlier decisions. This can aid in deciding how to implement change without violating that original intent of those decisions. And when required changes do violate that intent rationale can help in fixing problems by guiding the generation and selection of alternative means for satisfying that original intent.

Rationale from construction and use of the software can provide feedback that alerts requirements engineers, designers and managers to the need for change. In particular, user-centric rationale methods, such as Scenario-Claims Analysis and Case-Based Design Aids, can play a decisive role in detecting needs to changes in requirements, design and construction. Decision-centric rationale can also play a vital role in detecting the need for change by encouraging the participation of users and clients in the SE process. It does this by making the decision making processes of software engineers *transparent* to users and clients, i.e. open for inspection and evaluation. This tends to provoke responses from those users and clients, thus encouraging their participation. In fact, this use of decision-centric rationale was one of the main motivations for Rittel's pioneering work in design rationale (Rittel 1972).

Finally, rationale helps to manage change by documenting the intent of the changes themselves, so that these changes and their intent are not violated by future revisions. This is especially important when those making future changes are not the same people who made earlier changes. The importance of rationale in this case is due to the fact that changes are often made to decisions only after the initial, *intuitive* decisions failed to live up to expectations. The lesson about the failure of those earlier decisions results from hard-won experience and is, by definition, *counterintuitive*. So if those responsible for making future changes do not have access to the rationale for previous changes, they are very likely to "correct" them by restoring the decisions to their original, "intuitive" but erroneous states. In such cases, recording the rationale for changes is even more important than recording the rationale for the original, "intuitive" decisions.

18.4.2 Using Rationale to Manage the Increasing Scale, Complexity and Longevity of Software Projects

18.4.2.1 Using Rationale to Promote Coordination and Collaboration

Decision-centric rationale can play a role in promoting coordination and collaboration amongst the members of a project team. This works for the same reason that it works in facilitating participation, namely, that revealing the reasoning behind a decision enables others to critically evaluate that reasoning and thus participate intelligently in the decision making. Ultimately, collaboration and rationale management are mutually beneficial and interdependent, because 1) communicating rationale is the basis for collaboration and 2) collaborative communication is the best available source of rationale.

18.4.2.2 Using Rationale for Managing Large and Diverse Project Teams

If extensive amounts of rationale are generated by all the members of a project team, then managers of the project have a crucial means for monitoring all aspects of the development effort. It can become clear when projects are slipping behind schedule and in what areas they are slipping. Potential conflicts of decisions can be spotted before too much work is invested in building on or implementing flawed or inconsistent decisions. The need for additional collaboration may become clear to managers before it is clear to the potential collaborators themselves.

18.5 Challenges for Rationale-Based Software Engineering

For the potential of Rationale-Based Software Engineering to be realized there are a number of research challenges that must be met. One is that researchers need to continue to explore the role of rationale in SE activities that go beyond design and requirements engineering. Another is that much more research is needed on methods and systems for rationale management in support of iterative approaches to development, including incremental, evolutionary, and agile development as well as Extreme Programming. Such research is still in its infancy. Meeting these challenges will require a great deal of work, but does not appear particular problematic.

By contrast there are two research challenges that have been known about for more than a decade but with respect to which there has until recently been little progress. These are the *capture problem* and the *delivery problem*.

18.5.1 Solving the Capture Problem

Rationale-Based Software Engineering offers considerable promise, but there is also a substantial amount of work that has to be done before it can fully realize its potential and live up to its promise. The problem that has proved to be by far the most challenging is the so-called *capture problem*. The name is somewhat misleading, because what most researchers mean when they talk about the capture problem is a collection of three things combined: eliciting rationale from decision makers, structuring that rationale—e.g. according to a given conceptual schema—and recording that rationale in structured form. People combine these three things because rationale capture has traditionally involved all three of these activities.

The thing that makes rationale capture hard to accomplish is the structuring. This process tends to be highly labor intensive and may actually disrupt design thinking (Fisher et al. 1996); so decision makers are often unwilling to do it. As a consequence, it often happens that little rationale gets captured.

18.5.1.1 Exploiting Unique Characteristics of Software Development to Help Solve the Capture Problem

Amid the dire warnings about the future of software development, one interesting piece of good news is that the larger and more diverse development teams that are likely to be increasingly common in future software development will be far better sources of the software engineering rationale that will aid them in performing their jobs. The reason for this is that larger and more diverse teams inevitably require more explicit communication to collaborate and to coordinate their activities. And informally stated rationale is a major part of all such communication. This means that there will be more rationale that is captured as a side-effect of the normal development processes. This in turn means that, *the rationale capture problem*, which has been the biggest problem facing rationale researchers, will be much easier to solve. The emphasis of research will need to be less on capturing rationale than on indexing it for retrieval. If this can be done effectively, it will be a relatively simple matter to build libraries of project rationale.

It should be remembered that software development differs from the development of almost all other artifacts in the fact that every parts of the software lifecycle takes place on the computer—including system use and all activities of software development. This means that rationale can be captured from communication amongst all stakeholders in the project, including users; it also means that this captured rationale can be structured by being linked to the structure of the artifact itself, i.e. the structure of the software. These factors reduce the work of eliciting and recording rationale as well as the work of structuring rationale.

18.5.1.2 Re-using Rationale to Alleviate the Capture Problem

The basic idea behind the re-use of rationale is that since it so much work to elicit, structure and record rationale from scratch, it would be good if software engineers could take advantage of the fact that other decision makers had already gone to the trouble of doing this for their own project rationale. That way software engineers could "copy, paste and edit" their rationale rather than having to think up all the rationale from scratch. This would enable them to save their energy and resources—such as time, money and manpower—for the parts of their project that were unique. In other words, they would re-use rationale for the same basic reasons that we seek to re-use code rather than code everything from scratch.

The simplest way of re-using rationale is in the case of re-design. There they can simply use the rationale for the previous version of the software that they are redesigning. This rationale would have to be altered, but it I would still be the rationale with the closest fit to the current task. The only problem with this strategy is that it presumes someone have already put in the work of eliciting, structuring and recording the relevant rationale. It would be good if one could start a new software project by using the rationale for another project as a starting point.

Despite the unique features of each project, there are often important commonalities between projects. Many of the same decisions tasks, decision alternatives and evaluation argumentation are often found in prior projects. Even if the constellation of factors is different for each project, there are still similar *parts* of the rationale. A previous project can help to make sure that the crucial topics are dealt with using the crucial information. So even if the solutions to two projects are quite different, there may still be crucial overlaps in the rationale used.

There are a number of ways of re-using rationale. One is to build a case library of prior projects. This is, in effect, a type of case-based reasoning. A fundamentally different approach to re-using rationale would be to use rationale associated with the re-usable information known as design patterns. A number of researchers have worked on this already and there is likely to be much more work in this area in the future. Yet another approach would be to use Domain-Oriented Issue Bases (DOIBs), such are used by the JANUS (Fischer et al. 1996) and PHIDIAS (McCall et al. 1992) systems. DOIBs feature the issues, positions, arguments and dependency relationships that commonly arise in the various project within a given application domain. While no one has attempted to build DOIBs using argumentative approaches other than PHI (McCall 1991), there seems to be no principled reason why this could not be done.

18.5.2 Solving the Delivery Problem

Better means are needed for getting rationale to those who need it when they need it. This might be done by improving the ways in which rationale is indexed for retrieval. It might also be done by integrating rationale retrieval into the various tools used for decision making throughout software development.

Improving retrieval ultimately cannot solve the entire rationale delivery problem adequately. Retrieval only works when someone thinks to search for the information in an information system. Often people do not realize that a system contains information that affects their decision making. So there need to ways for people to be alerted to the existence of information they need even when they do not know to ask for it. Some such alerting mechanisms have been developed for special situations (Fischer et al. 1996) (Burge and Brown 2006), but more mechanisms and more general mechanisms need to be developed.

In SE every decision making and usage experience takes place on the computer. And a high percentage of the communication amongst stakeholders in a project is mediated by computer. In addition to aiding rationale capture, this would also facilitate rationale delivery. If rationale that refers to a part or feature of the software being developed were actually linked to the part or feature, it would be easy to retrieve that rationale by using the software itself as an index to it. Ideally this would work best if both the source code and the running code could be used to retrieve the rationale in this manner. It should also be noted that the same piece of rationale might be linked to several different parts of code and at several different levels of grouping in the hierarchy of code features or parts.

Linking rationale to the software itself also provides a way of alerting members of the software team to the existence of newly created rationale that are relevant to their interests. Generally, each member's responsibility for software features/parts is clearly assigned. These assignments can be used to alert the members of the team to the existence of new rationale that affects the decisions that they are responsible for. This strategy is similar to the one used by PHIDIAS' knowledge-based agents to alert team members to potential opportunities for collaboration (McCall et al. 1997) as well as Burge's strategy in the SEURAT system (Burge and Brown 2006).

Another approach could use dependency relationships between decisions to alert team members to changes in decisions that their own decisions depend on. For the design decisions that implement certain requirements, this approach could be used to alert designer when those requirements are modified in any way, or even when the rationale for those requirements is modified. This approach could also alert programmers who implement certain design features about any change in those features or in the rationale for those features. A variation of this approach might alert designer to the addition of new requirements, and alert programmers to the addition of new features to the design of the system.

18.6. Summary and Conclusions

Rational-based software engineering has a great deal to offer software engineers to help them cope with emerging problems in software development. Realizing this potential will require improvements in the way rationale is captured and delivered to those who need it. But the unique features of software development and the progress made in rationale research make it clear that success is possible. In fact, because of the way in which software development differs from the development of other artifacts, software engineering is likely to succeed in using rationale management before any other field that seeks to design and construct artifacts.

Bibliography

Abbattista F, Lanubile F, Mastelloni G, Visaggio G (1994) An experiment on the effect of design recording on impact analysis. In: Müller A, Georges A (eds) Proceedings of the international Conference on Software Maintenance. IEEE Computer Society, pp 253-259

Agrawal H, Horgan JR (1990) Dynamic program slicing. In: Proceedings of the ACM SIGPLAN 1990 Conference on Programming Language Design and Implementation, White Plains, New York, pp 246-256

Alexander C, Ishikawa S, Silverstein M, King I, Angel S, Jacobson M (1977), A pattern language: Towns, buildings, construction, Oxford University Press

Alford, MW, Lawson JT (1979) Software Requirements Engineering Methodology (Development). RADC-TR-79-168, U.S. Air Force, Rome Air Development Center, Griffiss AFB, New York, N.Y.

Ali-Babar M, Gorton I, Kitchenham BA (2006) A framework for supporting architecture knowledge and rationale management. In: Dutoit AH, McCall R, Mistrík I, Paech B (eds) Rationale Management in Software Engineering. Springer, pp 237-254

Ali-Babar M, Gorton I (2007) A tool for managing software architecture knowledge, In: the Workshop on the Sharing and Reusing Architectural Knowledge at the International Conference on Software Engineering, Minnesota

Ambler SW (1998) Software Process Patterns, Cambridge University Press and Reasoning (KR2002). pp 375-384

Andréka H, Ryan M, Schobbens PY (2002) Operators and laws for combining preference relations. Journal of Logic Computation 12(1):13-53

ANSI/IEEE (1987) ANSI/IEEE Standard for Software Unit Testing, ANSI/IEEE Std 1008-1987

Antón AI, Dempster JH, Siege DF (2000) Deriving goals from a use case based requirements specification for an electronic commerce system. In: Proceedings of the Sixth International Workshop on Requirements Engineering: Foundation for Software Quality (REFSQ), Stockholm, Sweden, pp 10-19

Antón AI, Potts C (2001) Functional paleontology: system evolution as the user sees it. In: Proceedings of the 23rd international Conference on Software Engineering, Toronto, Canada, pp 421-430

Antón, AI, Potts C (1998) The use of goals to surface requirements for evolving systems. In: Proceedings of the 20th international Conference on Software Engineering. pp 157-166

Antoniol G, Canfora G, de Lucia A, Casazza G (2000) Information retrieval models for recovering traceability links between code and documentation. In: Proceedings of the international Conference on Software Maintenance. pp 40-51

Arrow KJ (1963) Social Choice and Individual Values, 2nd edn, Wiley, New York

Avellis G, Borzacchini L, Cavallo A, Cotugno P, De Mastro G (1993) A blackboard architecture for intelligent assistance in software maintenance. In: Proceedings of the 6th International Workshop on Computer-Aided Software Engineering, Singapore, pp 180-189

Baker ER (2001) Which way, SQA?. IEEE Software. 18(1):16-18

Ball L, Lambell N, Omerod T, Slavin S, Mariani J (1999) Representing design rationale to support innovative design reuse: a minimalist approach. In: Proceedings of the 4th Annual Design Research Thinking Symposium, pp I.75-I.87

Bañares-Alcántara R, King MP, Ballinger G (1995) Egide: a design support system for conceptual chemical process design. In: AI System Support for Conceptual Design: Proc. of the 1995 Lancaster International Workshop on Engineering Design, Springer-Verlag, New York, pp 138-152

Baniassad ELA, Murphy GC, Schwanninger C (2003) Design pattern rationale graphs: Linking design to source. In: Proceedings of the 25th International Conference on Software Engineering (ICSE 2005), May 3-10, pp 352-362

Bass L, Clements P, Kazman R (2003) Software Architecture in Practice, 2nd edn. Addison-Wesley

Bass L, Clements P, Nord RL, Stafford J (2006) Capturing and using rationale for a software architecture. In: Dutoit AH, McCall R, Mistrík I, Paech B (eds) Rationale Management in Software Engineering, Springer, pp 255-272

Beck K (1999) Extreme Programming Explained: Embrace Change, Addison-Wesley

Beck K (2002) Test-Driven Development: By Example, Addison-Wesley

Beck K, Andres C (2005) Extreme Programming Explained: Embrace Change, 2nd edn, Addison Wesley, Boston

Bennett KH and Rajlich VT (2000) Software maintenance and evolution: A roadmap. In: Finkelstein, A (ed) The Future of Software Engineering, 22nd ICSE. Limerick, Ireland, pp 73-87

Bertolino, A (2007) Software testing research: achievements, challenges, dreams. In: Proceedings of the 29th International Conference on Software Engineering, Minneapolis, Minnesota, pp 85-103

Biggerstaff TJ, Mitbander BG, Webster, D (1993) The concept assignment problem in program understanding. In: Proceedings of the 15th international Conference on Software Engineering. Baltimore, Maryland, pp 482-498

Blackorby C, Donaldson D, Mongin P (2000) Social aggregation without the expected-utility hypothesis. U.B.C. Department of Economics Discussion Paper No. 00-18

Blevins D (2001) Overview of the Enterprise JavaBeans component model. In: Councill B, Heineman G (eds) Component-based Software Engineering, Addison-Wesley, Upper Saddle River, pp 589-606

Boehm B (1979) Software engineering: R&D trends and defence needs. Research Direction in Software Technology, Wegner P (ed) MIT Press, Cambridge MA, pp 1-9

Boehm B (1986) A spiral model of software development and enhancement. In: ACM SIGSOFT Software Engineering Notes. 11(4):22-42

Boehm B (2006) A view of 20th and 21st century software engineering. In: Proceedings of the 28th international Conference on Software Engineering, Shanghai, China, pp 12-29

Boehm B, Bose P (1994) A collaborative spiral software process model based on Theory W. In: Proceedings of the 3rd International Conference on the Software Process, Reston, VA, pp 59-68

Boehm B, Bose P, Horowitz E, Lee MJ (1995) Software requirements negotiation and renegotiation aids. In: Proceedings of the 17th international Conference on Software Engineering. Seattle,Washington, pp 128-142

Boehm B, Brown J, Kaspar H, Lipow M, MacLeod G, Merrit M (1979) Characteristics of Software Quality. TRW Series of Software Technology vol 1. North-Holland

Boehm B, Egyed A (1998) Software requirements negotiation: some lessons learned. In: Proceedings of the 20th international Conference on Software Engineering. Kyoto, Japan, IEEE Computer Society, pp 503-506

Boehm B, In H (1996) Identifying quality-requirement conflicts. IEEE Software. 13(2):25-35

Boehm B, Kitapci H (2006) The WinWin approach: using a requirements negotiation tool for rationale capture and use. In: Dutoit A, McCall R, Mistrík I, Paech B (eds) Rationale Management in Software Engineering, Springer, pp 173-190

Boehm B, Ross R (1989) Theory W software project management: principles and examples. IEEE Transactions on Software Engineering. 15(7):902-916

Boehm BW, Brown JR, Lipow M (1976) Quantitative evaluation of software quality. In: Proceedings of the 2nd International Conference on Software Engineering. San Francisco, California, pp 592-605

Boehm B (2006) Value-based software engineering: overview and agenda. In: Biffl S, Arum A, Boehm B, Erdogmus H, Grunbacher P (eds) Value-Based Software Engineering, Springer, pp 3-14

Boehm Barry, Prasanta Bose, Ellis Horowitz and Ming June Lee (1995) Software requirements negotiation and renegotiation aids: A Theory-W based spiral approach. In: Proceedings ICSE-17 (International Conference on Software Engineering), pp 243-253 IEEE Computer Society

Bohner SA, Arnold RS (1996) An introduction to software change impact analysis. In: Bohner SA, ArnoldRS (eds) Software Change Impact Analysis. IEEE Computer Society Press, pp 1-28

Booth R (2002) Social contraction and belief negotiation. In: Proceedings of the Eighth International Conference on Principles of Knowledge Representation and Reasoning (KR2002), pp 375-384

Bosch J (2004) Software architecture: The next step. Proc. 1st European Workshop Software Architecture (EWSA 04), Springer, pp 194-199

Bose P (1995) A model for decision maintenance in the WinWin collaboration framework. In: Proceedings of the 10th Conference on Knowledge Based Software Engineering. Boston, MA, pp 105-113

Bose, P (1998) Change analysis in an architectural model: A design rationale based approach. In: Proceedings ISAW3 (International Software Architecture Workshop), Orlando, Florida, pp 5-8

Bozheva T, Gallo M (2006) Defining Agile Patterns. In: Dutoit A, McCall R, Mistrík I, Paech B (eds) Rationale Management in Software Engineering, Springer, pp 373-390

Brice A, Johns B (1998) Improving process design by improving the design process. QSL-9002A-WP-001. QuantSci

Brooks FP (1995) The mythical man-month. Anniversary Edition, 2nd edn Reading, MA: Addison-Wesley

Bruegge B, Dutoit AH (2004) Object-Oriented Software Engineering Using UML, Patterns, and Java. 2nd edn Prentice Hall, NJ

Bruegge, B, Dutoit, A (2000) Object-Oriented Software Engineering: Conquering Complex and Changing Systems, Prentice Hall

Buckingham Shum S (2007), Hypermedia discourse: Contesting networks of ideas and arguments, Keynote Address, In: Proceedings of 15th International Conference on Conceptual Structures, Sheffield, UK, July 2007, Lecture Notes in Computer Science. 4604:29-44.

Buckingham Shum S, Hammond N (1994) Argumentation-based design rationale: What use at what cost? International Journal of Human-Computer Studies, 40(4), 603-652

Buckingham Shum S, MacLean A, Bellotti VME, Hammond NV (1997) Graphical Argumentation and Design Cognition, Human-Computer Interaction. 12(3):267-300

Buckingham Shum S, Selvin AM, Sierhuis M, Conklin EJ, Haley CB, Nuseibeh B (2006) Hypermedia Support for Argumentation-Based Rationale: 15 Years on from gIBIS and QOC. In: AH, McCall R, Mistrik I, Paech B (eds) Rationale Management in Software Engineering, Springer-Verlag: Berlin, pp 111-132

Budgen, D (2003) Software design, 2nd edn Addison-Wesley, Harlow, England

Burge J, Brown D (2006) Rationale-based support for software maintenance. In: Moran TP, Carroll JM (eds) Design rationale: Concepts, techniques, and use. Lawrence Erlbaum, Mahwah, NJ, pp 273-296

Burge J, Brown DC (2002) Integrating design rationale with a process model, In: Workshop on Design Process Modeling. Artificial Intelligence in Design '02. Cambridge, UK

Burge JE (2005) Software Engineering Using design RATionale. Ph.D. thesis, Worcester Polytechnic Institute

Burge JE, Brown DC (2000) Inferencing over design rationale. In: Artificial Intelligence in Design '00, Gero J (ed) Kluwer Academic Publishers, pp 611-629

Burge JE, Brown DC (2006) Rationale-based support for software maintenance. In: Dutoit AH, McCall R, Mistrik I, Paech B (eds) Rationale Management in Software Engineering, Springer, Germany, pp 273-296

Burge JE, Brown DC (2007) Supporting requirements traceability with rationale, GTC'07: International Symposium on Grand Challenges in Traceability, March 2007, Slade, KY

Burge JE, Brown DC (2003) Rationale support for maintenance of large scale systems. In: Proceedings of the Workshop on Evolution of Large-Scale Industrial Software Applications (ELISA), ICSM '03, Amsterdam, Netherlands

Burge JE, Brown DC (2004) An integrated approach for software design checking using rationale. In: Design Computing and Cognition '04, Gero J (ed) Kluwer Academic Publishers, pp 557-576

Burge JE, Cross V, Kiper J, Maynard-Zhang P, Cornford S (2006) Enhanced design checking involving constraints, collaboration, and assumptions. In: Gero J (ed) Proceedings of the Conference on Design, Computing, and Cognition. Eindhoven Netherlands, pp 655-674

Burnstein I (2003) Practical Software Testing, Springer Professional Computing

Canfora G, Casazza G, De Lucia A (2000) A Design rationale based environment for cooperative maintenance. International Journal of Software Engineering and Knowledge Engineering 10(5):627-645

Canfora G, Cerulo L (2006) Fine grained indexing of software repositories to support impact analysis. In: Proceedings of the 2006 international Workshop on Mining Software Repositories, Shanghai, China, pp 105-111

Capilla R, Nava F, Duenas JC (2007) Modeling and documenting the evolution of architectural design decisions. In: the Proceedings of the Workshop on Sharing and Reusing Archtectural Knowledge at the International Conference of Software Engineering (ICSE), Minneapolis Minnesota

Capilla R, Nava F, Pérez S, Dueñas JC (2006) A web-based tool for managing architectural design decisions. In: Proceedings of the Workshop on Sharing and Reusing Architectural Knowledge. ACM Digital Library. Software Engineering Notes 31 (5)

Carroll J (2000) Making use: scenario-based design of human-computer interaction, MIT press, Cambridge, MA

Carroll J, Rosson MB (1996) Deliberated evolution: stalking the View Matcher in design space. In: Moran TP, Carroll JM (eds) Design rationale: Concepts, techniques, and use. Lawrence Erlbaum, Mahwah, NJ, pp 107-145

Carroll JM, Rosson M (1992) Getting around the task-artifact cycle: how to make claims and design by scenario. ACM Trans. Inf. Syst. 10(2):181-212

Carroll JM, Rosson MB, Chin G Jr, Koenemann J (1998) Requirements development in scenario-based design. IEEE Transactions on Software Engineering. 24(12):1156-1170

Carroll JM, Mack, RL (1985) Metaphor, computing systems, and active learning. International Journal of Man-Machine Studies, 22:39-58

Carroll JM (1995) Scenario-based design: Envisioning work and technology in system development. New York: John Wiley & Sons

Carroll JM, Alpert SR, Karat J, Van Deusen, MD, Rosson, MB (1994) Capturing design history and rationale in multimedia narratives. In: Proceedings of CHI'94: Human Factors in Computing Systems. Boston. New York: ACM Press/Addison-Wesley, pp 192-197

Carroll, JM, Rosson, MB, Convertino, G, Ganoe, C (2006) Awareness and teamwork in computer-supported collaborations. Interacting with Computers, 18: 21-46

Chaib-Draa B, Dignum, F (2002) Trends in agent communications language. Computational Intelligence, 18(2):89-101

Chapin N (2000) Software maintenance types—a fresh view. In: Proceedings of the International Conference on Software Maintenance, San Jose, CA, pp 247-252

Charette RN (1996) Large-scale project management is risk management. IEEE Software, 13(4):110-117

Chaudron MRV, Groote JF, van Hee KM, Hemerik C, Somers LJAM, Verhoeff T (2004) Software Engineering Reference Framework. Technical Report CS-Report 04-039, Computer Science Reports, Department of Mathematics and Computer Science, Eindhoven University of Technology, Eindhoven, The Netherlands

Chernak Y (2001) Validating and Improving Test-Case Effectiveness. IEEE Software 18(1):81-86

Chewar CM, Bachetti E, McCrickard DS, Booker J (2005) Automating a design reuse facility with critical parameters: lessons learned in developing the LINK-UP System. In: Jacob R, Limbourg Q, Vanderderonckt J (eds) Computer-Aided Design of User Interfaces IV. Kluwer Academic Publishers, pp 235-246

Chung L, Nixon B, Yu E (1996) Dealing with change: an approach using non-functional requirements. Requirements Eng 1(4):238-260

Chung L, Nixon BA (1995) Dealing with non functional requirements: three experimental studies of a process-oriented approach. In: Proceedings of the 17th International Conference on Software Engineering, pp 25-37

Chung L, Nixon BA, Yu E, Mylopoulos J (2000) Non-functional requirements in software engineering, Kluwer Academic Publishers.

Chung L, Yu E (1998) Achieving system-wide architectural qualities. In: Proceedings of the OMG-DARPAMCC Workshop on Compositional Software Architectures

Chung PWH, Goodwin R (1998) An integrated approach to representing and accessing design rationale, Engineering Applications of Artificial Intelligence, 11:149-159

Cimitile A, Lanubile F, Visaggio G (1992) Traceability based on design decisions. In: Proceedings of the Conference on Software Maintenance, Orlando, Florida, pp 309-317

Clapp J (1993) Getting started on software metrics. IEEE Software. 10(1):189-109, 117

Clark A (1987) From folk psychology to naïve psychology. Cognitive Science, 11, 139-154

Cleland-Huang J, Settimi R, BenKhadra O, Berezhan E, Christina S (2005) Goal centric traceability for managing non-functional requirements. In: Proceedings of the International Conference on Software Engineering, St Louis, pp 362-371

Clemen RT, Reilly T (2001) Making Hard Decisions. Duxbury, Forest Grove, CA

Clemen RT, Winkler RL (1999) Combining probability distributions from experts in risk analysis, Risk Analysis 19:187-203

Clements P, Bachmann, Bass L, Garlan, Ivers J, Little R, Nord R, Stafford J (2002) Documenting software architecture: Views and beyond, Addison-Wesley

Clements P, Northrop L (2002) Software Product Lines Practices and Patterns, Addison-Wesley

CMMI Product Team (2006) CMMI For Development. Version 1.2. CMU/SEI-2006-TR-008

CMU (2002) Quality measures taxonomy http://www.sei.cmu.edu/str/taxonomies/view_qm.html

Coleman JS (1990) The foundations of social theory, Cambridge, MA: Harvard University Press

Conklin EJ, Burgess-Yakemovic KC (1996) A process-oriented approach to design rationale. In: Moran T, Carroll J (eds) Design Rationale Concepts, Techniques, and Use. Lawrence Erlbaum Associates, pp 393-427

Conklin J, Begeman M (1988) gIBIS: A hypertext tool for exploratory policy discussion, ACM Transactions on Office Information Systems, 6(4):303-331

Conklin J, Burgess-Yakemovic K (1995) A process-oriented approach to design rationale. In: Design rationale concepts, techniques, and use, Moran T, Carroll J (eds). Lawrence Erlbaum Associates, pp 293-428

Conklin Jeff, Burgess-Yakemovic K (1991) A process-oriented approach to design rationale, human-computer interaction, 6:357-291

Connolly T, Jessup L, Valacich J (1990) Effects of anonymity and evaluative tone on idea generation in computer-mediated groups. Management Science, 36(6): 689-703

Coplien JO (1995) A generative development-process pattern language. In: Coplien JO, Schmidt DC (eds) Pattern Languages of Program Design, Addison Wesley Longman, Inc, pp 183-237

Councill B, Heineman G (2001) Definition of a software component and its elements. In: Councill B, Heineman G (eds) Component-based Software Engineering, Addison-Wesley, Upper Saddle River, pp 5-20

Crosby ME, Scholtz J, Wiedenbeck S (2002) The roles beacons play in comprehension for novice and expert programmers. In: Kuljis K, Baldwin L, Scoble R (eds) Proceedings of the Psychology of Programming Interest Group, Brunel University, pp 58-73

Cross N (2003) Evidence from protocol and other formal studies of design activity. In: Eastman C, McCracken M, Newstetter W (eds), Knowing and Learning to Design: Cognitive Perspectives in Design Education, Amsterdam: Elsevier

Curtis B, Krasner H, Iscoe N (1988) A field study of the software design process for large systems. Commun. ACM 31(11):1268-1287

Cysneiros LM, Leite JCSP (2001) Using UML to reflect non-functional requirements. In: Proceedings of the 11th CASCON, November, IBM Canada, Toronto, pp 202-216

Cysneiros LM, Leite JCSP (2004) Nonfunctional requirements: from elicitation to conceptual models. IEEE Transactions on Software Engineering, 30(5):328-350

Darimont R, Delor E, Massonet P, van Lamsweerde A (1997) GRAIL/KAOS: an environment for goal-driven requirements engineering. In: Proceedings of the 19th international Conference on Software Engineering. ACM Press, New York, NY, pp 612-613

de Boer RC, Farenhorst R, Clerc V, van der Ven JS, Lago P, van Vliet, H (2006) A model for structuring software architecture project memories, In: Proceedings of the 8th International Workshop on Learning Software Organizations

De Grace P, Stahl LH (1998) Wicked problems, righteous solutions: a catalogue of software engineering paradigms, Yourdon Press

de Kleer J (1986) An assumption-based truth maintenance system. Artificial Intelligence 28(2): 127-162

de la Garza J, Alcantara P (1997) Using parameter dependency network to represent design rationale. Journal of Computing in Civil Engineering, 2(2): 102-112

Dellen B, Kohler K, Maurer F (1996) Integrating software process models and design rationales. In: Proceedings of 11th Knowledge-Based Software Engineering Conference (KBSE'96) September 25-28, Syracuse, NY, pp 84-93

DeMarco T, Lister TR (1999) Peopleware: Productive projects and teams, 2nd edn, New York: Dorset House Publishing

Demeyer S, Ducasse S, Nierstrasz O (2003) Object-Oriented Reengineering Patterns, Morgan Kaufmann Publishers

Department of Defense (2002) Mandatory Procedures for Major Defense Acquisition Programs (MDAPS) and Major Automated Information System (MAIS), DoD 5000.2-R

Devanbu P, Brachman R, Selfridge P, Ballard B (1991) Lassie: a knowledge-based software information system. In: Communications of the ACM, 34(5):34-49

Dick J (2005) Design traceability. IEEE Software. 22(6):14-16

Dijkstra EW (1972) The humble programmer. Communications of the ACM 15(10): 859-866

Dijsktra EW (1989) On the cruelty of really teaching computer science. Communications of the ACM, 32(12):1398-1404

Do H, Rothermel G, Kinneer A (2006) Prioritizing JUnit test cases: An empirical assessment and cost-benefits analysis. Empirical Software Engineering, 11: 33-70

Domeshek E, Kolodner JL (1996) The Designers' Muse: Providing Experience to Aid Conceptual Design of Complex Artifacts. In Maher, M.L. and Pu, P. (eds), Issues and Applications of Case-Based Reasoning to Design, Lawrence Erlbaum Associates: Mahwah, NJ, pp 11-38

Doyle J (1979) A truth maintenance system. Artificial Intelligence 12(3): 231-272

Druffel Larry E, Buxton John N (October 1980) Requirements for an Ada programming support environment: Rationale for Stoneman, In: Proceedings COMPSAC Chicago, pp 66-72

Dutoit AH, Paech B (2000) Supporting Evolution: Using Rationale in Use Case Driven Software Development, Proceedings of the Sixth International Workshop on Requirements Engineering: Foundation for Software Quality (REFSQ'2000), Stockholm, Sweden

Dutoit A, McCall R, Mistrik I, Paech B (eds) (2006) Rationale management in software engineering, Springer Verlag, Heidelberg

Dutoit A, McCall R, Mistrík I, Paech B (2006) Rationale management in software engineering: Concepts and techniques. In: Dutoit A, McCall R, Mistrík I, Paech B (eds) Rationale management in software engineering, Springer-Verlag, pp 1-48

Egyed A, Grunbacher P (2004) Identifying requirements conflicts and cooperation: How quality attributes and automated traceability can help. IEEE Software. 21(6):50-58

EIA (1998) Electronic Industries Alliance. Systems Engineering Capability Model (EIA/IS-731). Washington, DC

Eick S (1998) Maintenance of large systems. In: Stasko J, Dominigue J, Brown M, Price B (eds) Software Visualization: Programming as a Multimedia Experience. The MIT Press, pp 315-328

Eickelmann NS, Ruffolo F, Baik J, Anant A (2002) An Empirical Study of Modifying the Fagan Inspection Process and the Resulting Main Effects and Interaction Effects Among Defects Found, Effort Required, Rate of Preparation and Inspection, Number of Team Members and Product 1st Pass Quality. In: Proceedings of the 27th Annual NASA Goddard Software Engineering Workshop (Sew-27'02), pp 58-64

Erdogmus H, Favaro J, Halling M (2006) Valuation of software initiatives under uncertainty: concepts, issues, and techniques. In: Biffl S, Aurum A, Boehm B, Erdogmus H, Grunbacher P (eds) Value-Based Software Engineering, Springer, pp 39-66

Ewald T (2001) Overview of COM+, In: Councill B, Heineman G (eds) Component-based Software Engineering, Addison-Wesley, Upper Saddle River, pp 573-588

Fabian A, Wahid S, Bhatia S, McCrickard DS (2006) Creating an Interactive Learning Environment with Reusable HCI Knowledge. In: Proceedings of the World Conference on Educational Multimedia/Hypermedia and Educational Telecommunications (ED-MEDIA '06), Orlando FL, June, pp 2314-2322

Falessi D, Beker M, Cantone G (2006) Design decision rationale: experiences and steps ahead towards systematic use. In: Proceedings of the Workshop on the Sharing and Reuse of Architectural Knowledge, Torino, Italy

Favaro J (1996) When the pursuit of quality destroys value. IEEE Software. 13(3):93-95

Filman RE (1998) Achieving ilities. In: Proceedings of the Workshop on Compositional Software Architectures, Monterey, CA, USA

Finkelstein A, Kramer J (2000) Software engineering: a roadmap. In: Proceedings of the Conference on the Future of Software Engineering, ICSE 2000, Limerick, Ireland, pp 3-22

Finkelstein AC, Gabbay D, Hunter A, Kramer J, Nuseibeh, B (1994) Inconsistency handling in multiperspective specifications. IEEE Transactions on Software Engineering 20(8):569-578

Fischer G and Morch A (1988) CRACK: A critiquing approach to cooperative kitchen design. In: Proceedings of the international conference on intelligent tutoring systems (Montreal, Canada), ACM Press, New York, pp 176-185

Fischer G, Henninger S, Redmiles D (1991) Cognitive tools for locating and comprehending software objects for reuse. In: Proceedings of the 13th international Conference on Software Engineering, Austin, Texas, pp 318-328

Fischer G, Lemke A, McCall R, Morch A (1995) Making argumentation serve design. In: Moran T, Carroll J (eds) Design rationale concepts, techniques, and use, Lawrence Erlbaum Associates, pp 267-294

Fischer G, McCall R, Morch A (1989) JANUS: Integrating hypertext with a knowledge-based design. In: Proceedings of Hypertext '89, pp 105-117

Fisher R, Ury W (1981) Getting to Yes. Houghton-Mifflin

Fitzpatrick (2003) The locales framework: understanding and designing for wicked problems Fitzpatrick G, Series: Computer Supported Cooperative Work , vol 1, Springer Verlag, New York

Fowler M, Beck K, Brant J, Opdyke W, Roberts D (1999) Refactoring: Improving the Design of Existing Code, Addison-Wesley

Fowler M (2003) UML distilled, 3rd edn, New York: Addison-Wesley

Fox J, Das S (2000) Safe and Sound Artificial Intelligence in Hazardous Applications, AAAI Press

France RB, Ghosh S, Dinh-Trong T, Solberg A (2006) Model-Driven Development Using UML 2.0: Promises and Pitfalls. Computer 39(2):59-66

Frankl P, Hamlet D, Littlewood B, Strigini L (1997) Choosing a testing method to deliver reliability. In: Proceedings of the 19th international Conference on Software Engineering, Boston, Massachusetts, pp 68-78

Fuggetta A (2000) Software process: a roadmap. In: Proceedings of the Conference on the Future of Software Engineering, Limerick, Ireland, pp 25-34

Gallagher KB, Lyle JR (1991) Using program slicing in software maintenance. In: IEEE Transactions on Software Engineering 17(8):751-761

Gamma E, Helm R, Johnson R, Vlissides J (1995) Design patterns: Elements of reusable object-oriented software. Addison-Wesley, Reading, MA

Ganeshan R, Garrett J, Finger S (1994) A framework for representing design intent, In: Design Studies, 15(1):59-84

Gannod G, Burge J, Urban S (2007) Issues in the design of flexible and dynamic service-oriented systems. In: Proceedings of the International Workshop on Systems Development in SOA Environments. Co-located with ICSE, Minneapolis, MN.

Gibson VR, Senn JA (1989) System structure and software maintenance performance. Communications of the ACM. 32(3):347-358

Gilboa I, Samet D, Schmeidler D (2004) Utilitarian aggregation of beliefs and tastes. Journal of Political Economy, 112(4):932-938

Godfrey M, Tu Q (2002) Tracking structural evolution using origin analysis. In: Proceedings of the international Workshop on Principles of Software Evolution, Orlando, Florida, pp 117-119

Goldenson DR, Gibson DL (2003) Demonstrating the impact and benefits of CMMI: an update and preliminary results. CMU/SEI-2003-SR-009

Gomaa H, Farrukh GA (1998) Composition of software architectures from reusable architecture patterns. In: Proceedings of the 3rd International Workshop on Software Architecture, Orlando, Florida, pp 45-48

Gotel O, Finkelstein A (1994) An analysis of the requirements traceability problem. In: Proceedings of the 1st Int. Conf. on Requirements Engineering, IEEE Computer Society Press, pp 94-101

Gruber TR and Russell DM (1996) Generative design rationale: beyond the record and reply paradigm, Design rationale; concepts, techniques and use, Lawrence Erlbaum Associates, Mahwah, NJ, pp 323-349

Grudin J (1988) Why CSCW applications fail: Problems in the design and evaluation of organizational interfaces. Proceedings of the CSCW '88 Conference on Computer-Supported Cooperative Work, ACM, New York, pp 85-93

Grudin J (1996) Evaluating opportunities for design capture. In: Moran T, Carroll J (eds) Design rationale concepts, techniques, and use, Lawrence Erlbaum

Grudin J (1994) Groupware and social dynamics: Eight challenges for developers. Communications of the ACM, 37(1):92-105

Grünbacher P, Boehm B (2001) EasyWinWin: a groupware-supported methodology for requirements negotiation. In: Proceedings of the 8th European Software Engineering Conference.Vienna, Austria, ACM Press, pp 320-321

Grundy J, Hosking J, Mugridge R (1998) Inconsistency management for multiple-view software development environments. In: IEEE Transactions on Software Engineering. 24(11):960-981

Guindon R (1990) Knowledge Exploited by Experts During Software System Design, I. J. Man-Machine Studies 33: 279-304

Hagge L, Houdek F, Lappe K, Paech B (2006) Using patterns for sharing requirements engineering process rationales. In: Dutoit A, McCall R, Mistrík I, Paech B (eds) Rationale Management in Software Engineering, Springer, pp 409-426

Hall T, Rainer A, Baddoo N, Beecham S (2001) An empirical study of maintenance issues within process improvement programmes in the software industry. In: Proc. of the International Conference on Software Maintenance. Florence, Italy, pp 422-430

Harrold MJ (2000) Testing: a roadmap. In: Proceedings of the Conference on the Future of Software Engineering. Limerick, Ireland, pp 61-72

Harsanyi J (1955) Cardinal welfare, individualistic ethids, and interpersonal coparisions of utility. Journal of Political Economy 63:309-321

Hayes JH, Dekhtyar A, Sundaram SK (2005) Improving after-the-fact tracing and mapping: supporting software quality predictions. IEEE Software. 22(6): 30-37

Haynes, S (2006) Three studies of design rationale as explanation, Dutoit A, McCall R, Mistrík I, Paech B (eds) Rationale management in software engineering. Springer Verlag, Heidelberg, pp 53-71

Heinemann GT, Council WT (2001) Component-based Software Engineering. Addison-Wesley, Upper Saddle River

Herbsleb JD (1999) Metaphorical representation in collaborative software engineering. ACM Work Activities Coordination and Collaboration Conference: WACC '99. San Francisco, New York: ACM, pp 117-126

Hild M, Jeffrey R, Risse M (1998) Preference aggregation after Harsanyi. In: Salles M, Weymark J (eds) Justice, Political Liberalism, and Utilitarianism. Cambridge: Cambridge University Press

Hirsch M (2002) Making RUP agile. In: OOPSLA 2002 Practitioners Reports, Seattle, Washington, ACM Press, New York, NY

Hofmann HF, Lehner F (2001) Requirements Engineering as a Success Factor in Software Projects. IEEE Software. 18(4):58-66

Hordijk W, Wieringa R (2006) Reusable rationale blocks: improving quality and efficiency of design choices. In: Dutoit A, McCall R, Mistrík I, Paech B (eds) Rationale management in software engineering, Springer, pp 353-371

Houdek F, Kempter H (1997) Quality patterns—an approach to packaging software engineering experience. In: Proceedings of the 1997 symposium on Software reusability. Boston, Massachusetts, pp 81-88

Hudlicka E (1997) Summary of knowledge elicitation techniques for requirements analysis. Course Material for Human Computer Interaction, Worcester Polytechnic Institute

Hull MC, Jackson K, Dick J (2002) Requirements Engineering. Springer-Verlag. London, UK

Humphrey W (1995) A Discipline for Software Engineering, Addison-Wesley, Reading, MA

Humphrey W (2000) The Personal Software Process. CMU/SEI-2000-TR-022

IEEE (1990) IEEE standard glossary of software engineering terminology. IEEE Std 610.12-1990

IEEE (1993) IEEE Std 610.12-1990. IEEE Standard Glossary of Software Engineering Terminology. IEEE

IEEE (1998) IEEE Standard for Software Maintenance. IEEE Std 1219-1998

IEEE (2000) IEEE Std 1471-2000. Recommended Practice for Architectural Descriptions of Software-Intensive Systems, IEEE

IEEE (2004) Guide to the Software Engineering Body of Knowledge (SWEBOK) 2004 edn, IEEE

IEEE (2004) IEEE standard for software verification and validation. IEEE Std 1012-2004

IEEE/EIA (1996) Software life cycle processes. ISO/IEC 12207.0-1996

ISO (1999) Requirements for Enterprise Reference Architecture and Methodologies, ISO TC 184/SC5/WG1

Jacobson I, Booch G, Rumbaugh J (1999) The Unified Software Development Process, Addison-Wesley Publishing Co

Jackson M (2007) Specialization in Software Engineering. Keynote paper. To appear in: Proceedings of 14th Asia Pacific Software Engineering Conference (APSEC 2007), Nagoya, Japan, 5-7 December 2007, IEEE Computer Society Press.

Jensen RW, Tonies CC (1979) Software Engineering. Prentice Hall, NJ

Jung H, Kim S, Chung C (2004) Measuring software product quality: a survey of ISO/IEC 9126. IEEE Software, 21(5):88-92

Juristo N, Moreno AM, Strigel W (2006) Guest Editors' Introduction: Software Testing Practices in Industry. IEEE Software. 23(4):19-21

Kahn H (1962) Thinking about the unthinkable, New York: Horizon Press

Kahneman D, Tversky A (2000) Choices, values, and frames, New York: Cambridge University Press

Kajiko-Mattsson M (2001) The state of documentation practice within corrective maintenance. In: Proceedings of the International Conference on Software Maintenance, pp 354-363

Karacapilidis N, Papadias D (2001) Computer supported argumentation and collaborative decision making: the HERMES system, Information Systems, 26(4): 259-277

Karat J, Carroll JM, Alpert SR, Rosson, MB (1995) Evaluating a multimedia history system as support for collaborative design. In: Nordby K, Helmersen P, Gilmore D, Arnesen S (eds) Proceedings of Human-Computer Interaction — INTERACT'95 Lillehammer, Norway, London: Chapman & Hall, pp 346-353

Karau SJ, Williams KD (1993) Social loafing: A meta-analytic review and theoretical integration. Journal of Personality & Social Psychology, 65(4): 681-706

Karlsson J, Ryan K (1997) A cost-value approach to prioritizing requirements. IEEE Software. 14(5):67-74

Kazman Rick, Asundi Jai, Klein Mark (2003) Quantifying the value of architecture design decisions: lessons from the field, Proceedings ICSE-25 (International Conference on Software Engineering), Portland, OR: IEEE Computer Society, pp 557-562

Kelly T, Littman J (2001) Lessons in Creativity from IDEO, America's Leading Design Firm. New York: Doubleday & Company

Kimelman D, Rosenburg B, Roth T (1998) Visualization of dynamics in real world software systems. In: Stasko J, Dominigue J, Brown M, Price B (eds) Software Visualization: Programming as a Multimedia Experience. The MIT Press, pp 293-314

King JMP, Bañares-Alcántara R (1997) Extending the scope and use of design rationale records. In: Artificial Intelligence for Engineering Design, Analysis and Manufacturing. 11:155-167

Kirschner, PA, Buckingham Shum, SJ, Carr, CS (eds) (2003) Visualizing argumentation. London: Springer

Kitchenham B, Linkman S (1998) Validation, verification, and testing: Diversity rules. IEEE Software. 15(4):46-49

Kitchenham BA, Travassos GH, von Mayhauser A, Niessink F, Schneidewind NF, Singer J, Takada S, Vehvilainen R, Yang, H (1999) Towards an ontology of software maintenance, Journal of Software Maintenance: Research and Practice, 11:365-389

Klein M (1997) An exception handling approach to enhancing consistency, completeness and correctness in collaborative requirements capture. Concurrent Engineering Research and Applications 5(1):37-46

Klein M, Kazman R (1999) Attribute-Based Architectural Styles. CMU/SEI-99-TR-022. Software Engineering Institute, Carnegie Mellon University, Pittsburg, PA

Kleppe J, Warmer J, Bast W (2003) MDA explained, the model driven architecture: practice and promise, Addison-Wesley

Knodel J, Muthig D (2006) The role of rationale in the design of product line architectures – a case study from industry. In: Dutoit A, McCall R, Mistrík I, Paech B (eds) Rationale Management in Software Engineering, Springer, pp 297-312

Knuth DE (1992) Literate programming. Stanford, California: Center for the Study of Language and Information, CSLI Lecture Notes, no. 27

Ko AJ, Aung H, Myers BA (2005) Eliciting design requirements for maintenance-oriented IDEs: a detailed study of corrective and perfective maintenance tasks. In: Proceedings of the 27th international Conference on Software Engineering. St. Louis, MO, pp 126-135

Kolodner J (1993) Case-based reasoning. Morgan Kaufmann Publishers, San Mateo, California

Kontio J (1996) A case study in applying a systematic method for COTS selection In: Proceedings of the 18th International Conference on Software Engineering. Berlin, Germany, pp 201-209

Koschke R, Quante J (2005) On dynamic feature location. In: Proceedings of the 20th IEEE/ACM international Conference on Automated Software Engineering. Long Beach, CA, pp 86-95

Kraut RE, Streeter LA (1995) Coordination in software development, Communications of the ACM, 38(3), 69-81

Kraut RE (2003) Applying social psychological theory to the problems of group work. In J.M. Carroll (ed), HCI Models, theories, and frameworks: Toward a multidisciplinary science. San Francisco: Morgan-Kaufmann, pp 325-356

Kruchten P (1999) The Rational Unified Process: An Introduction, Addison-Wesley

Kruchten P, Lago P, van Vliet H (2006) Building up and reasoning about architectural knowledge. In: Proceedings of the Second International Conference on the Quality of Software Architectures (QoSA 2006)

Kunz W, Rittel HWJ (1970) Issues as elements of information systems, Working Paper 131, Center for Urban and Regional Development, University of California, Berkeley

Kuusela J, Savolainen J (2000) Requirements engineering for product families. In: Proceedings of the 22nd international Conference on Software Engineering, Limerick, Ireland, pp 61-69

Lai VS, Wong BK, Cheung W (2002) Group decision making in a multiple criteria environment: A case using the AHP in software selection, In: European Journal of Operational Research 137:134-144

Lam W, Shankararaman V (1999) Requirements change: a dissection of management issues. In: Proceedings of the 25th Euromicro Conference, pp 2244

Lammers S (1986) Programmers at Work. Microsoft Press, Redmond, Washington.

Lanza M (2001) The evolution matrix: recovering software evolution using software visualization techniques. In: Proceedings of the 4th international Workshop on Principles of Software Evolution. Vienna, Austria, pp 37-42

Larman C (2004) Agile and Iterative Development: A Manager's Guide, Addison-Wesley, New York

Larman C, Basili VR (2003) Iterative and incremental development: A brief history, IEEE Computer, IEEE, pp 47-56

Latane B, Bourgeois, MJ (2001) Dynamic social impact and the consolidation, clustering, correlation, and continuing diversity of culture. In: Hogg MA, Tindale RS (eds) Blackwell handbook of social psychology: Group processes. Oxford, UK: Blackwell

Lave J, Wenger E (1991) Situated Learning. Legitimate peripheral participation, Cambridge, UK: University of Cambridge Press

Lave J (1988) Cognition in practice: Mind, mathematics and culture in everyday life, New York: Cambridge University Press

Law J, Rothermel G (2003) Whole program Path-Based dynamic impact analysis. In: Proceedings of the 25th international Conference on Software Engineering, Portland, Oregon, pp 308-318

Lawson B (1979) Cognitive Strategies in Architectural Design. Ergonomics 22(1): 59-68

Leavitt HJ (1951) Some effects of certain communication patterns on group performance. Journal of Abnormal and Social Psychology, 46:38-50.

Lee J (1990) SIBYL: A tool for managing group decision rationale. Proceedings of the 1990 ACM Conference on Computer-Supported Cooperative Work, ACM, New York, pp 79-92

Lee J (1991) Extending the Potts and Bruns model for recording design rationale. In: Proceedings of the 13th International Conference on Software Engineering, Austin, TX, pp 114-125

Lee J, Lai KY (1995) What's in design rationale? In: Moran T, Carroll J (eds) Design rationale concepts, techniques, and use, Lawrence Erlbaum Associates, pp 21-51

Lee J (1990) SIBYL: A qualitative design management system, In: Winston PH, Shellard, S (eds) Artificial Intelligence at MIT: Expanding frontiers, pp 104-133, Cambridge MA: MIT Press

Lehman MM (1996) Laws of software evolution revisited. In: Montangero C (ed.) Lecture Notes In Computer Science, vol. 1149. Springer-Verlag, London, pp 108-124

Lehman MM (2005) The role and impact of assumptions in software development, maintenance, and evolution. In: Proceedings of the IEEE International Workshop on Software Evolvability. Budapest, Hungary, pp 3-14

Lehman MM, Frenández-Ramil J (2006) The role and impact of assumptions in software engineering and its products. In: Dutoit AH, McCall R, Mistrik I, Paech B (eds) Rationale Management in Software Engineering. Springer, Germany, pp 311-328

Lewis C, Rieman J, Bells B (1996) Problem-centered design for expressiveness. In: Moran TP, Carroll JM (eds) Design rationale, concepts, techniques and use, Lawrence Erlbaum Associates, Mahwah, New Jersey, pp 147-184

Lewis JA, Henry SA, Kafura DG, Schulman RS (1991) An empirical study of the Object-Oriented paradigm and software reuse. In: Proceedings of OOPSLA, pp 184-196

Lientz BP, Swanson, EB (1988) Software maintenance management. Addison-Wesley

Lindquist, C (2005) Fixing the requirements mess. CIO Magazine. 15 November, http://www.cio.com/archive/111505/require.html

Liskov, Guttag (1986) Abstraction and specification in program development, MIT Press, Cambridge

Lloyd P, Scott P (1994) Discovering the Design Problem. Design Studies 15(2): 125-140

Lougher R, Rodden T (1993) Group support for the recording and sharing of maintenance rationale. Software Engineering Journal 8(6):295-306

Lozano-Tello, Adolfo, Asuncin Gomez-Perez (2002) BAREMO: How to choose the appropriate software component using the analytic hierarchy process. In: Proceedings of the 14th International Conference on Software Engineering and Knowledge Engineering, pp 781-788

Mackay WE, Ratzer AV, Janecek P (2000) Video artifacts for design: Bridging the gap between abstraction and detail. In: ACM DIS 2000: Conference on Designing Interactive Systems. Brooklyn, New York, New York: ACM, pp 72-82

MacLean A, Young RM, Bellotti VME, Moran T (1996) Questions, Options and Criteria. In: Moran TP, Carroll JM (eds) Design rationale, concepts, techniques and use, Lawrence Erlbaum Associates, Mahwah, New Jersey, pp 53-106

MacLean A, Young RM, Moran TP (1989) Design rationale: the argument behind the artifact. In: Proceedings of the SIGCHI conference on human factors in computing systems: wings for the mind. ACM Press, pp 247-252

Madsen KH (1994) A guide to metaphorical design. Communications of the ACM, 37(12):57-62

Maletic JI, Marcus A (2001) Supporting program comprehension using semantic and structural information. In: Proceedings of the 23rd international Conference on Software Engineering. Toronto, Canada, pp 103-112

Mannion M, Keepence B, Kaindl H, Wheadon J (1999) Reusing single system requirements from application family requirements. In: Proceedings of the

International Conference on Software Engineering. Limerick, Ireland, pp 453-462

Manola F (1999) Providing Systematic Properties (Ilities) and Quality of Service in Component-Based Systems. Technical report, Object Services and Consulting, Inc

Martin J (1991) Rapid Application Development. Macmillan Publishing Company, NY

Matena V, Hapner M (1999) Enterprise JavaBeans specification. v1.1. Sun Microsystems, java.sun.com/products/ejb/docs.html

Mayer B (1997) Object-oriented software construction, 2nd edition, Prentice-Hall

Maynard-Ried II P, Shoham Y (2001) Belief fusion: aggregating pedigreed belief states. Journal of Logic, Language and Information. Kluwer Academic Publishers. 10:183-209

Maynard-Zhang P, Lehman D (2003) Representing and aggregating conflicting beliefs. Journal of Artificial Intelligence Research 19:155-203

McAndrews DR (2000) The Team Software Process (TSPSM): An Overview and Preliminary Results of Using Disciplined Practices. Technical Report CMU/SEI-2000-TR-015

McCall R (1979) Final Report for Project STIEC (Scientific and Technical Information in the European Community), Studiengruppe fuer Systemforschung, Heidelberg

McCall R (1979) On the structure and use of issue systems in design, Doctoral Dissertation 1978, University of California, Berkeley, University Microfilms

McCall R (1986) Issue-Serve Systems: A Descriptive Theory of Design, In: Design Methods and Theories, vol 20, no 3, DMG, San Luis Obispo, pp 443-458

McCall R (1991) PHI: a conceptual foundation for design hypermedia. Design Studies 1:30-41

McCall R and Johnson E (1997) Using argumentative agents to catalyze and support collaboration in design, Automation in Construction, Elsevier Science BV, 6(4):299-309

McCall R, Bennett P, d'Oronzio P, Ostwald J, Shipman F, Wallace N (1990) Phidias: A PHI-Based Design Environment Integrating CAD Graphics into Dynamic Hypertext. In: Proceedings of the European Conference on Hypertext (ECHT 1990)

McCall R, Bennett P, d'Oronzio P, Oswald J, Shipman FM III, Wallace N (1992) PHIDIAS: Integrating CAD graphics into dynamic hypertext. In: Streitz N, Rizk A, André J (eds) Hypertext: Concepts, Systems and Applications, Cambridge University Press, New York, NY, pp 152-165

McCall R, Johnson E (1997) Using argumentative agents to catalyze and support collaboration in design, Automation in Construction, 6(4):299-309

McCall R, Mistrik I, Schuler W (1981) An Integrated Information and Communication System for Problem Solving. Proceedings of the Seventh International CODATA Conference. Pergamon, London

McCall R, Mistrik I (2005) Capture of software requirements and rationale through collaborative software development. Requirements Engineering for

Sociotechnical Systems, Mate JL and Silva A (eds) Information Science Publishing, Hershey, PA, pp 303-317

McCall RJ (1989) MIKROPLIS: a hypertext system for design. Design Studies 10(4):228-238

McKerlie D, MacLean A (1993) QOC in action (abstract): using design rationale to support design. In: Proceedings of the SIGCHI Conference on Human Factors in Computing Systems. Amsterdam, The Netherlands ACM Press, New York, NY, pp 519

Mehta A, Heineman GT (2002) Evolving legacy system features into fine-grained components. In: Proceedings of the 24th International Conference on Software Engineering. Orlando, Florida, pp 417-427

Mellor SJ, Clark AN, Futagami T (2003) Guest editors' introduction: Model-driven development. IEEE Software. 20(5):14-18

Mens K, Mens T, Wermelinger M (2002) Supporting software evolution with intentional software views. In: Proceedings of the International Workshop on Principles of Software Evolution. Orlando, Florida, pp 138-142

Mens T, Demeyer S (2001) Future trends in software evolution metrics. In: Proceedings of the 4th International Workshop on Principles of Software Evolution, Vienna, Austria, pp 83-86

Meyer T, Ghose AK, Chopra S (2001) Multi-agent context-based merging. In: Proceedings of Common Sense 2001: The Fifth Symposium on Logical Formalizations of Commonsense Reasoning, New York, USA, May 2001

Mockus A, Fielding RT, Herbsleb JD (2002) Two Case Studies of Open Source Software Development: Apache and Mozilla. ACM Transactions on Software Engineering and Methodology. 11(3):209-346

Mohamed A, Ruhe G, Eberlein A (2005) Decision support for customization of the COTS selection process. In: Proceedings of 2nd International Workshop on Models and Processes for the Evaluation of COTS Components (MPEC'05). 27th International Conference on Software Engineering. St. Louis, Missouri, pp 1-4

Mongin, P (1998) The paradox of the Bayesian experts and state-dependent utility theory. Journal of Mathematical Economics 29(3):331-361

Moran TP, Carroll JM (1996) Design rationale: concepts, techniques and use, Lawrence Erlbaum Associates, Mahwah, New Jersey

Moreira A, Rashid A, Araujo B (2005) Multi-dimensional separation of concerns in requirements engineering. In: Proceedings of the 13th International Conference on Requirements Engineering Paris, France, pp 285-296

Morisio M, Seaman C, Parra A, Basili VR, Condon S, Kraft S (2000) Investigating and improving a COTS-Based software development process. In: Proceedings of the 22nd International Conference on Software Engineering. Limerick, Ireland, pp 32-41

Myer B (1988) Object Oriented Software Construction, Prentice Hall

Myers KL, Zumel NB, Garcia PE (1999) Automated capture of rationale for the detailed design process. In: Proceedings of the Eleventh National Conference on Innovative Applications of Artificial Intelligence (IAAI-99), AAAI Press, Menlo Park, CA, pp 876-883

Narayanan NH and Kolodner JL (1995) Case libraries in support of design education: the DesignMuse Experiences, Proceedings of the 25th Annual Frontiers in Education Conference, IEEE Press

Nentwich C, Emmerich W, Finkelstein A, Ellmer E (2003) Flexible consistency checking. ACM Transactions on Software Engineering Methodology 12(1): 28-63

Newman MW, Landay, JA (2000) Sitemaps, Storyboards, and Specifications: A Sketch of Web Site Design Practice. In Proceedings of Designing Interactive Systems: DIS 2000. New York, NY. pp 263-274, August 17-19, 2000

Ng TH, Cheung SC, Chan WK, Yu YT (2006) Toward effective deployment of design patterns for software extension: a case study. In: Proceedings of the 2006 international Workshop on Software Quality. Shanghai, China, pp 51-56

Ngo-The A, Ruhe G (2005) Decision Support in Requirements Engineering, In: Arum A, Wohlin C (eds) Engineering and Managing Software Requirements. Springer, pp 267-286

Nierstrasz O, Ducasse S, Gïrba T (2005) The story of moose: an agile reengineering environment. In: Proceedings of the 10th European Software Engineering Conference Held Jointly with 13th ACM SIGSOFT international Symposium on Foundations of Software Engineering. Lisbon, Portugal, pp 1-10

Nuseibeh B (1997) Ariane 5 Who Dunnit?, IEEE Software, 14(3):15-16

Nuseibeh B, Easterbrook S (2000) Requirements engineering: a roadmap. In: Proceedings of the Conference on the Future of Software Engineering. Limerick, Ireland, pp 35-46

Nuseibeh B, Easterbrook S, Russo A (2000) Leveraging Inconsistency in Software Development. Computer 33(4):24-29

Oberto R (2002) FAIR/DART Option #2. Advanced Projects Design Team. NASA Jet Propulsion Laboratory

Object Management Group (2000) The Common Object Request Broker: Architecture and Specification Version 2.4

Oinas-Kukkonen H (1988) Evaluating the usefulness of design rationale in CASE, European Journal of Information Systems, 7(3):185-191

OMG (2005a) UML Object Constraint Language (OCL) Specification. v2.0. The Object Modeling Group

OMG (2005b) UML Superstructure. v2.1.1, The Object Modeling Group

Ossher H, Tarr P (1999) Multi-dimensional Separation of Concerns in Hyperspace, Tech. Report RC 21453(96717), IBM, T.J. Watson Research Center, NY

Overhage S (2004) UnSCom: A standardized framework for the specification of software components. In: Weske M, Liggesmeyer P (eds) Object-Oriented and Internet-Based Technologies, 5th Annual International Conference on Object-Oriented and Internet-Based Technologies, Concepts, and Applications for a Networked World. Lecture Notes in Computer Science (LNCS) 3263, Springer, pp 169-184

Parnas DL, Clements PC (1986) A Rational Design Process: How and why to fake it. IEEE Transactions on Software Engineering, SE-12(2):251-257

Parsons S (2001) Qualitative Methods for Reasoning under Uncertainty, MIT Press

Parsons S, Hunter A (1998) A review of uncertainty handling formalisms. In: Hunter A, Parsons S (eds) Applications of Uncertainty Formalisms. Lecture Notes In Computer Science, vol 1455. Springer-Verlag, London, pp 8-37

Patterson DA (2005) 20th Century vs. 21st Century C&C: The SPUR Manifesto, The President's Letter, Communications of the ACM, Vol 48, No 3, pp 15-16

Paulk MC, Curtis B, Chrissis MB, Weber CV (1993) Capability Maturity Model for Software, Version 1.1. CMU/SEI-93-TR-024

Payne C, Allgood CF, Chewar CM, Holbrook C, McCrickard DS (2003) Generalizing Interface Design Knowledge: Lessons Learned from Developing a Claims Library. IEEE International Conference on Information Reuse and Integration (IRI 2003), Las Vegas, pp 362-369

Pelled L, Eisenhardt K, Xin K (1999) Exploring the black box: An analysis of work group diversity, conflict, and performance. Administrative Science Quarterly, 44:1-28

Peña-Mora F, Sriram D, Logcher R (1995) Design rationale for computer-supported conflict mitigation. ASCE Journal of Computing in Civil Engineering 9(1):57-72

Peña-Mora F, Vadhavkar S (1997) Augmenting design patterns with design rationale. Artificial Intelligence for Engineering Design, Analysis and Manufacturing 11(2):93–108

Pennock D, Maynard-Ried II P, Giles CL, Horvitz E (2000) A normative examination of ensemble learning algorithms. In: Proceedings of the 17th International Conference on Machine Learning, pp 735-742

Perl J (1988) Probabilistic Reasoning in Intelligent Systems: Networks of Plausible Inference, Morgan Kaufmann

Pigoski TM (1998) Practical Software Maintenance. John Wiley & Sons

PMBOK (2003) A Guide to the Project Management Body of Knowledge. Project Management Institute Standards Committee, IEEE Std 1490-2003

Pollice G, Augustine L, Lowe C, Madhur J (2003) Software Development for Small Teams: a Rup-Centric Approach. Addison Wesley Longman Publishing Co., Inc

Potts C (1996) Supporting software design: integrating design methods and design rationale. In: Moran TP, Carroll JM (eds) Design rationale: concepts, techniques, and use. Lawrence Erlbaum Associates, Mahwah, New Jersey, pp 295-321

Potts C, Bruns (1988) Recording the reasons for design decisions, In: Proceedings of the 10th international conference on software engineering, Singapore, pp 418-427

Potts, Colin (1989) A generic model for representing design methods, Proc. 11th Int. Conf. Software Eng., Pittsburgh, IEEE Comp. Soc. Press

Potts Colin, Bruns Glen (1988) Recording the reasons for design decisions. In: Proceedings of the 10th International Conference on Software Engineering, Singapore, IEEE Comp. Soc. Press, pp 418-427

Potts Colin, Takahashi Kenji, Anton Anna (March1994) Inquiry-based requirements analysis, IEEE Software. 11(2):21-32

Prechelt L, Unger B, Tichy WF, Brössler P, Votta LG (2001) A controlled experiment in maintenance comparing design patterns to simpler solutions. IEEE Trans. Softw. Eng. 27(12):1134-1144

Price B, Baecker R, Small I (1998) An introduction to software visualization. In: Stasko J, Dominigue J, Brown M, Price B (eds) Software Visualization: Programming as a Multimedia Experience. The MIT Press, pp 3-27

Queille J, Voidrot J, Wilde N, Munro M (1994) The impact analysis task in software maintenance: a model and a case study. In: Proceedings of the international Conference on Software Maintenance, Victoria, BC, pp 234-242

Rajlich V (2006) Changing the software paradigm, Communications of the ACM, 49(8):67-70

Ramesh B, Dhar V (1992) Supporting systems development by capturing deliberations during requirements engineering, IEEE Transactions on Software Engineering, 18(6): 498-510

Ramesh B, Dhar V (1994) Representing and maintaining process knowledge for large-scale systems development, EEE Expert. 9(2):54-60

Ramires J, Antunes P, Respício A (2005) Software requirements negotiation using the software quality function deployment. In: Fuks H, Lukosch S, Salgado A (eds) Groupware: Design, Implementation, and Use. Lecture Notes in Computer Science. Vol 3706, Heidelberg, Springer-Verlag, pp 308-324

Ramler R, Biffl S, Grunbacher P (2006) Value-based management of software testing. In: Biffl S, Aurum A, Boehm B, Erdogmus H, Grunbacher P (eds) Value-Based Software Engineering, Springer

Randal B, Buxton JN (1970) Software engineering techniques: A report on a conference sponsored by the NATO Science Committee, NATO

Rasmussen J (1974) The human data processor as a system component: Bits and pieces of a model. Riso-M-1722. Roskilde, Denmark: Danish Atomic Energy Commission

Raymond E (2001) The Cathedral and the Bazaar. Revised Edition. O'Reilly

Redwine S, Riddle W (1985). Software technology maturation, In: Proc. of the 8th ICSE, pp 189-200

Reeves B, Shipman FM III (1992) Supporting communication between designers with artifact-centered evolving information spaces In: Proceedings of the 1992 ACM Conference on Computer-supported Cooperative Work, November 1-4, Toronto, Ontario, Canada, pp 394-401

Reiss SP (2002) Constraining software evolution. In: Proceedings of the International conference on Software maintenance, Montreal, Quebec Canada, pp 162-171

Reiss SP, Kennedy CM, Wooldridge T, Krishnamurthi S (2003) CLIME: an environment for constrained evolution demonstration description. In: Proceedings of the 25th international Conference on Software Engineering. Portland, Oregon, pp 818-819

Reitman WR (1965) Cognition and thought.: An information processing approach. New York: John Wiley and Sons

Réquilé-Romanczuk A, Cechich A, Dourgnon-Hanoune A, Mielnik J (2005) Towards a knowledge-based framework for COTS component identification. In: Proceedings of the Second international Workshop on Models and Processes for the Evaluation of off-the-Shelf Components. St. Louis, Missouri, ACM Press, New York, NY, pp 1-4

Riesbeck C, Schank R (1989) Inside Case-Based Reasoning. Lawrence Erlbaum. Hillsdale, NJ

Rittel H (1980) APIS: A concept for an argumentative planning information system. Working paper 324, Institute of Urban and Regional Development, University of California, Berkeley

Rittel H, Weber M (1973) Dilemmas in a general theory of planning. Policy Sciences 4: 155-169

Rittel HWJ (1972) On the planning crisis: Systems analysis of the first and second generations. Bedriftsokonomen, Norway, 8:390-396

Rittel, H. (1972) Son of Rittelthink, DMG Newsletter 3-10 Berkeley: University of California

Robillard MP (2005) Automatic generation of suggestions for program investigation. In: Proceedings of the 10th European Software Engineering Conference Held Jointly with 13th ACM SIGSOFT international Symposium on Foundations of Software Engineering. Lisbon, Portugal, pp 11-20

Roman G (1985) A taxonomy of current issues in requirements engineering. Computer, pp 14-22

Rosson MB, Alpert SA (1990) The cognitive consequences of object-oriented design. Human-Computer Interaction, 5, 345-379

Rosson MB, Carroll JM (1996) The reuse of uses in Smalltalk programming. ACM Transactions on Computer-Human Interaction, 3(3), 219-253

Rothermel G, Elbaum S (2003) Putting Your Best Tests Forward. IEEE Software. 20(5):74-77

Royce WW (1970) Managing the development of large software systems: concepts and techniques. In: Proceedings of IEEE WESTCON, Los Angeles, California, pp 1-9

Saaty TL (1990) Multicriteria Decision Making: The Analytic Hierarchy Process, RSW Publications

Saaty TL (1990) The Analytic Hierarchy Process, McGraw-Hill, New York

Saiedian H, Dale R (2000) Requirements engineering: making the connection between the software developer and customer, Information and Software Technology, 42:419-428

Schmid K (2002) A comprehensive product line scoping approach and its validation. In: Proceedings of ICSE. Orlando, Florida, pp 593-603

Schmidt DC (2006) Guest editor's introduction. Model-Driven Engineering. Computer, 39(2):25-31

Schmidt DC, Buschmann F (2003) Patterns, frameworks, and middleware: their synergistic relationships. In: Proceedings of the 25th international Conference on Software Engineering. Portland, Oregon, IEEE Computer Society, Washington, DC, pp 694-704

Schneider, K (2006) Rationale as a by-product, in Rationale management in software engineering, Dutoit, A McCall, R Mistrik, I Paech, B (eds) Springer Verlag, Heidelberg, pp 91-109

Schön D (1983) The reflective practitioner. How professionals think in action. Basic Books, New York

Schultz DJ (1979) A case study in system integration using the Build approach. In: Proceedings of the 1979 Annual Conference. pp 143-151

Scriven M (1967) The methodology of evaluation. In: Tyler R, Gagne R, Scriven M (eds) Perspectives of Curriculum Evaluation, Rand McNally, pp 39-83

SEI (1997) Integrated Product Development Capability Maturity Model, Draft Version 0.98. Pittsburgh, PA: Enterprise Process Improvement Collaboration and Software Engineering Institute, Carnegie Mellon University

Shafer G (1976) A Mathematical Theory of Evidence. Princeton University Press, Princeton, New Jersey

Sharp H, Finkelstein A, Galal G (1999) Stakeholder Identification in the Requirements Engineering Process. In: Proceedings of the 10th International Workshop on Database & Expert Systems Applications, pp 387

Shipman FM III, McCall R (1994) Supporting knowledge-base evolution with incremental formalization. In: Proceedings of the SIGCHI Conference on Human Factors in Computing Systems, Boston, Massachusetts, United States, pp 285-291

Shipman F, McCall R (1997) Integrating Different Perspectives on Design Rationale: "Supporting the Emergence of Design Rationale from Design Communication", Artificial Intelligence in Engineering Design, Analysis, and Manufacturing (AIEDAM), 11(2):141-154

Shipman FM III and Marshall CC (1999) Formality Considered Harmful: Experiences, Emerging Themes, and Directions on the Use of Formal Representations in Interactive Systems, Computer-Supported Cooperative Work, 8(3): 333-352

Shipman FM III, Marshall CC (1999) Spatial Hypertext: An Alternative to Navigational and Semantic Links, ACM Computing Surveys 31(4):1-5

Siff M, Reps T (1999) Identifying modules via concept analysis. IEEE Transactions on Software Engineering. 25(6):749-768

Sim S, Duffy A (1994) A new perspective to design intent and design rationale. In: Workshop Notes for Representing and Using Design Rationale. Artificial Intelligence in Design. pp 4-12

Simon HA (1957) Models of man: Social and rational. New York: Wiley

Singer J (1998) Practices of software maintenance. In: Proc. of the International Conference on Software Maintenance (ICSM'98). Bethesda, Maryland, USA, pp 139-145

Smith RP, Tjandra P (1998) Experimental Observation of Iteration in Engineering Design, Research in Engineering Design 10(2):107-117

Sneed H (1995) Planning the reengineering of legacy systems. IEEE Software. 12(1):24-34

Sneed HM (2001) Impact analysis of maintenance tasks for a distributed object-oriented system. In: Proceedings of the IEEE International Conference on Software Maintenance, Florence, Italy, pp 180-189

Software Engineering Coordinating Committee (2004) Guide to the Software Engineering Body of Knowledge (SWEBOK), IEEE Computer Society

Sommerville I (2007) Software Engineering. 8th edn, Addison Wesley

Spanoudakis G, Zisman A (2001) Inconsistency management in software engineering: survey and open research issues. In: Handbook of Software Engineering and Knowledge Engineering 1, World Scientific Publishing, pp 329-380

Srivastava A, Thiagarajan J (2002) Effectively prioritizing tests in development environment. In: Proceedings of the 2002 ACM SIGSOFT International Symposium on Software Testing and Analysis. Roma, Italy, pp 97-106

Stahl T, Volter M (2006) Model-driven software development, Wiley

Stark GE, Oman P, Skillicorn A, Ameele A (1999) An examination of the effects of requirements changes on software maintenance releases. Journal of Software Maintenance 11(5):293-309

Stobie K (2005) Too darned big to test. Queue 3(1):30-37

Stone M (1961) The linear opinion pool. Annals of Mathematics and Statistics 32:1339-1342

Subramaniam GV (2000) Object model resurrection — an object oriented maintenance activity. In: Proceedings of the 22nd international Conference on Software Engineering. Limerick, Ireland, pp 324-333

Subramanian N, Chung L (2001) Software architecture adaptability: an NFR approach. In: Proceedings of the 4th international Workshop on Principles of Software Evolution. Vienna, Austria, pp 52-61

Subramanian N, Chung L (2002) Tool support for engineering adaptability into software architecture. In: Proceedings of the international Workshop on Principles of Software Evolution. Orlando, Florida, pp 86-96

Sutcliffe AG, Carroll JM (1999) Designing Claims for Reuse in Interactive Systems Design. International Journal of Human-Computer Studies, 50(3):213-241

Tang A, Han J (2005) Architecture rationalization: A methodology for architecture verifiability, traceability and completeness. Proc. 12th Annual IEEE Int. Conf. and Workshop on the Engineering of Computer Based Systems (ECBS'05), pp 135-144

Tang A, Jin Y, Han J (2005) A rationale-based architecture model for design traceability and reasoning. Journal of Systems and Software, 80, pp 918-934

Tang A, Jin Y, Han J, Nicholson A (2005) Predicting change impact in architecture design with Bayesian Belief Networks. In: proceedings of the 5th Working IEEE/IFIP Conference on Software Architecture, pp 67-76

Tang A, Babar M, Gorton I, Han J, (2006) A survey of architecture design rationale. Journal of Systems and Software, 79:1792-1804

Tarr P, Clarke LA (1998) Consistency management for complex applications. In: Proceedings of the 20th international Conference on Software Engineering, Kyoto, Japan, pp 230-239

Taylor RN, van der Hoek A (2007) Software Design and Architecture: The once and future focus of software engineering. In: 2007 Future of Software Engineering. International Conference on Software Engineering. Minneapolis, MN, pp 226-243

Thayer RH, Dorfman M (1990) Introduction, issues, and terminology. In: Thayer RH, Dorfman M (eds) System and Software Requirements Engineering, IEEE Computer Society Press, Los Alamitos, CA, 1st edition, pp 1-3

Thelin T, Runeson P, Wohlin C (2003) An experimental comparison of usage-based and checklist-based reading. IEEE Transactions on Software Engineering. 29(8)687-702.

Tip F (1995) A survey of program slicing techniques. Journal of Programming Languages. 3(3):121-189

Tonella P, Antoniol G (1999) Object oriented design pattern inference. In: Proceedings of the IEEE international Conference on Software Maintenance, pp 230-238

Toulmin S (1958) The Uses of Argument. Cambridge University Press, Cambridge

Turner M, Budgen D, Brereton P (2003) Turning software into a service. IEEE Computer 36(10):38-44

Tyree J, Akerman A (2005) Architecture decisions: Demystifying architecture. IEEE Software 22(2):19-27

Ullman D (2004) An example of decision management. White Paper, http://www.robustdecisions.com/ decision-management.pdf

van Lamsweerde A (2001) Goal-oriented requirements engineering: a guided tour. In: Proceedings of the 5th IEEE International Symposium on Requirements Engineering, Toronto, pp 249-263

van Lamsweerde A (2003) From system goals to software architecture. In: Bernardo M, Inverardi P (eds) Formal Methods for Software Architectures, Springer-Verlag

van Lamsweerde A (2004) Goal-oriented requirements engineering: a roundtrip from research to practice. In: Proceedings of the 12th International Conference on Requirements Engineering, Tokyo, Japan, pp 4-8

van Lamsweerde A, Letier E (2000) Handling obstacles in goal-oriented requirements engineering. IEEE Transactions Software Engineering Special Issue on Exception Handling. 26(10): 978-1005

Vetschera R (2006) Preference-based decision support in software engineering. In: Biffl S, Aurum A, Boehm B, Erdogmus H, Grunbacher P (eds) Value-Based Software Engineering, Springer, pp 67-89

Vincenti W (1990) What engineers know and how they know it. John Hopkins University Press

Voas JM, Miller KW (1995) Software Testability: the new verification. IEEE Software 12(3):17-28

von Mayrhauser A, Vans AM (1994) Comprehension processes during large scale maintenance. In: Proceedings of the 16th international Conference on Software Engineering. Sorrento, Italy, pp 39-48

Wahid S, Smith J.L, Berry B, Chewar CM, McCrickard, DS (2004) Visualization of Design Knowledge Component Relationships to Facilitate Reuse. Proceedings of the 2004 IEEE International Conference on Information Reuse and Integration (IRI'04), Las Vegas NV, November, pp 414-419.

Walker RJ, Holmes R, Hedgeland I, Kapur P, Smith A (2006) A lightweight approach to technical risk estimation via probabilistic impact analysis. In: Proceedings of the 2006 international Workshop on Mining Software Repositories, Shanghai, China, pp 98-104

Wallin P, Fröberg A, Axelsson A (2007) Making decisions in integration of automotive software and electronics: A method based on ATAM and AHP, In: Proceedings SEAS'07 (Fourth International Workshop on Software Engineering for Automotive Systems), IEEE Computer Society, pp 5

Wang L, Tan KC (2005) Software testing for safety-critical applications. IEEE Instrumentation and Measurement Magazine. 8(2):38-47

Wang N, Schmidt D, O'Ryan C (2001) Overview of the CORBA component model. In: Councill B, Heineman G (eds) Component-based Software Engineering, Addison-Wesley, Upper Saddle River, pp 557-571

Wang Xin, Xiong Guangleng (2001) Design rationale as part of corporate technical memory, In: International Conference on Systems, Man, and Cybernetics, Tucson, AZ: 3:1904-1908

Weick KE (1995) Sensemaking in organizations. Sage Publications: Thousand Oaks, CA

Weiser M (1981) Program slicing. In: Proceedings of the 5th international Conference on Software Engineering, San Diego, California, pp 439-449

Weizenbaum, J (1966) ELIZA—A Computer Program For the Study of Natural Language Communication Between Man and Machine, Communications of the ACM 9(1):36-35

Wellman MP (1990) Fundamental concepts of qualitative probabilistic networks. Artificial Intelligence, 44:257-303

Wenger E (1998) Communities of practice: Learning, meaning, and identity. New York: Cambridge University Press

Whitehead J (2007) Collaboration in Software Engineering: A Roadmap. In: 2007 Future of Software Engineering, International Conference on Software Engineering. Minneapolis, MN, pp 214-225

Whittaker S, Schwarz H (1995) Back to the Future: Pen and Paper Technology Supports Complex Group Coordination, in Proceedings of ACM CHI'95 Conference on Human Factors in Computing Systems, pp 495-502

Workshop on Multi-Dimensional Separation of Concerns, International Conference on Software Engineering 2000, http://www.research.ibm.com/hyperspace/workshops/icse2000

Zadeh LA (1965) Fuzzy sets, Information and Control 8:338-353

Zadeh LA (1978) Fuzzy sets as a basis for a theory of possibility, Fuzzy Sets and Systems 1:3-28

Zhu L, Gorton I (2007) UML profiles for design decisions and non-functional requirements. In: Proceedings of the Workshop on the Sharing and Reusing of

Architectural Knowledge, at the International Conference of Software Engineering (ICSE), Minneapolis, Minnesota.

Zimmermann T, Weisgerber P, Diehl S, Zeller A (2004) Mining version histories to guide software changes. In: Proceedings of the 26th International Conference on Software Engineering. pp 563-572

Zimring CM, Bafna S, Do E (1996) Structuring cases in a case-based design aid. In: Vanegas J, Chinowsky P (eds), Third Congress on Design Computing, Anaheim, A/C/E'96, American Society of Civil Engineers, 17-19, June, 1996, pp 308-313

Zimring CM, Do ED, Domeshek ED, Kolodner JL (1995) Supporting case-study use in design education: A computational case-based design aid for architecture. In: Mohsen JP (ed) Computing in Civil Engineering: Proceedings of the Second Congress. New York: American Society of Civil Engineers, pp 1635-1642

Ziv H, Richardson DJ, Klosch R (1996) The uncertainty principle in software engineering. Technical Report UCI-TR-96-33, University of California, Irvine

Glossary

abstraction
A view of an object that focuses on the information relevant to a particular purpose and ignores the remainder of the information [IEEE Std 610.12-1990]

adaptability
Adaptability concerns the ease of altering a system to meet the needs of a user [Randal and Buxton 1970]

adaptive maintenance
Changes made to the software during maintenance that do not change its functionality

agile methods
Software development methods that use iterative development to provide a more agile response to changing requirements

analysis
The phase in the software life-cycle that analyzes the system requirements in order to build a model describing the application domain

analysis of design
A process that provides a *view* of the design process that is not otherwise available

Analytic Hierarchy Process (AHP)
A decision-making technique where alternatives are evaluated by making a series of pair-wise comparisons

anthropomorphism
Software analysis and design method that involves metaphorically thinking about software components as animate

anti-model
Vivid characterizations of features and outcomes that a problem solver or decision makers definitely wants to avoid

architectural description
A collection of products to document an architecture [IEEE Std 1471-2000]

architectural design
The result of the process of defining a collection of hardware and software components and their interfaces to establish the framework for the development of a computer system [adapted from IEEE Std 610.12-1990]]

architectural framework
Describes the elements of a concrete architecture in terms of components, connectors, and dependencies

architectural style
A family of architectures constrained by component/connector vocabulary, topology, semantic constraints [adapted from Garlan and Shaw 1993]

architectural tactic
A transformation of an architecture to achieve particular quality attribute goals

architecturally significant requirements (ASR)
Software requirements that have broad cross-functional implications such performance, usability, maintainability, security. These include Non-functional Requirements (NFRs) and quality attributes

architecture
Architecture is the fundamental organization of a system embodied in its components, their relationships to each other, and to the environment, and the principles guiding its design and evolution [IEEE Std 1471-2000].

Architecture is a description (model) of the basic arrangement and connectivity of parts of a system (either a physical or a conceptual object or entity) [ISO 15704 1999]

architecture decision
A high level design decision that an architect or designer takes to satisfy the functional and non-functional requirements of a system

architecture description
A collection of products to document an architecture [IEEE Std 1471-2000]

artifact
The result of any activity in the software life-cycle such as requirements, architecture model, design specifications, source code and test scripts [http://w3.umh.ac.be/genlog/SE/SE-contents.html]

assumption
A proposition that is believed to be, but not known to be, true

Attribute-based Architectural Styles (ABASs)
Architectural styles associated with an attribute reasoning framework associated with a quality attribute

awareness (in collaboration)

Relevant knowledge about collaborators; for example, their identity, activity, goals and expectations, focus of attention, and so forth. Effective collaboration requires awareness

bad smell

Code structures that signal potential problems or poor designs that indicate the need for refactoring

basic software

Software that is used by the computer hardware to give the system its basic functions, like an operating system, or performs elementary tasks such as a compiler [Chaudron et al. 2004]

benchmark

A benchmark is a set of tests used to compare the performance of alternative tools, methods, or techniques [http://w3.umh.ac.be/genlog/SE/SE-contents.html]

black-box reuse

A kind of reuse where a component is reused without changing anything within the component

black-box testing

Software testing that looks only at the inputs and expected outputs and is not aware of the internal contents of the code

business case

Description of the system in terms of the stakeholders that have to make the decision to develop the system, together with an analysis of the development and operational cost of the system, and of the benefits of the system and the revenues it might generate [Chaudron et al. 2004]

Capability Maturity Model (CMM)

A process model developed by the Software Engineering Institute to assess the maturity of a software organizations process by classifying it into one of five levels

Capability Maturity Model Integration (CMMI)

A replacement for the CMM that assesses the maturity level for twenty two process areas

CASE (Computer-Aided Software Engineering)

CASE is the use of software tools to assist in the development and maintenance of software

classical decision model

Decision model in which solution alternatives are exhaustively enumerated, analyzed, and contrastively evaluated

common ground

The knowledge shared by interlocutors or collaborators. Effective collaboration requires periodic verification of common ground

commonalities

The set of features or properties of a component (or system) that are the same, or common, between systems [http://w3.umh.ac.be/genlog/SE/SE-contents.html]

community of practice

Groups of actors that share values, norms, concepts, behavior scripts, and strategies pertaining to a domain of human endeavor

compatibility

The ability of two or more systems or components to perform their required function while sharing the same hardware or software environment [IEEE Std 610.12-1990]

component

A component is a self-contained piece of software with clearly-defined interfaces and explicitly-declared context dependencies [Stahl and Volter 2006]

Component Based Software Engineering (CBSE)

A software engineering approach that builds software systems from reusable software components

conceptual framework (or frame of reference)

Establishes terms and concepts pertaining to the content and use of a specific architectural descriptions

confirmation bias

Tendency of human decision makers to seek and prize data that confirms their decisions over data that disconfirms their decisions

consistency

The degree of uniformity, standardization, and freedom from contradiction among the documents or parts of a system or component [IEEE Std 610.12-1990]

consistency management

The process of managing the consistency between the different software artifacts developed during the software development process

coordination

Self-management among collaborators to ensure that individual contributions can be synthesized into effective wholes

corrective maintenance

Software maintenance changes that are made in order to repair defects

correctness

The ability of software products to perform their exact tasks, as defined by their specification [Meyer 1997]

COTS (Commercial Off-the-Shelf)

Software products that are purchased rather than custom made

criterion-based evaluation

A way of evaluating a decision alternative or artifact feature which consists of 1) the statement of a criterion, e.g. a goal, and 2) an assessment of the alternative or feature with respect to the stated criterion, these two elements in effect constituting a single argument for or against the alternative or feature

decision-centric rationale approaches

Rationale approaches that deal with rationale for the decision making in artifact creation

decision representation language (DRL)

Lee's revision and extension of the Potts and Bruns approach to rationale. DRL's schema corresponds roughly to a superset of QOC's schema and has dependency relationships between elements. Like QOC, DRL uses a form of criterion-based evaluation

defect

A problem in a software artifact that causes it to be incorrect

deliberate *(verb)*

To consider what the answer to *a question* should be and, more specifically, to evaluate one or more proposed answers to a question

design *(noun)*

An artifact description that is detailed enough to be used to construct that artifact

design pattern

Names, abstracts, and identifies the key aspects of a common design structure that make it useful for creating a reusable object-oriented design [Gamma et al. 1995]

design rationale (DR)

DR is the reasoning underlying the creation and use of artifacts

design space decisions

Decisions as to what features that an artifact will have

Design Space Analysis

Representation of a set of design space decision tasks together with their decision alternatives and the evaluations of these alternatives

domain-oriented issue base (DOIB)

PHI-based collections of issues, positions, arguments and sub-issues that commonly arise in a particular design domain

detailed design

The result of the process of refining and expanding the preliminary design of a system or component to the extent that the design is sufficiently complete to be implemented [IEEE Std 610.12-1990]

domain

An area of knowledge or activity characterized by a set of concepts and terminology understood by practitioners in that area [Booch et al. 1990]

enhancive maintenance

Software maintenance changes that are made to add additional features or otherwise improve a software system

extensibility

The ease of adapting software products to changes of specification [IEEE Std 610.12-1990]

extreme programming (XP)

A popular agile method that proposes taking software best-practices "to the extreme"

familiarity bias

Tendency of human decision makers to consider familiar data and interpretations as typical

fixation

Tendency of designers to make solution decisions before adequately understanding the full problem space, and then to disproportionate adduce confirmatory evidence to justify and maintain those decisions

flexibility

The case with which a system or component can be modified for use in application or environments other than those for which it was specifically designed [IEEE Std 610.12-1990]

framework

A framework is a generic structure that can be adapted or extended via systematic extension or configuration [adapted from Stahl and Volter, 2006]

functional design

The result of the process of defining the working relationships among the components of a system [IEEE Std 610.12-1990]

functional requirement

A requirement that describes functionality that the system must provide in order to be acceptable to the customer

functionality

The extent of services provided by a system [adapted from Mayer 1997]

generality

The degree to which a system or component performs a broad range of functions [IEEE Std 610.12-1990]

generative paradigm

An alternative devised by Gruber and Russell to the "record and replay" paradigm used by almost all other approaches to rationale. Rather than re-cording rationale, the generative paradigm involves re-creating recreating it after-the-fact by deriving it from various data obtained automatically during design

glass-box testing

Software testing that is based on information about the structure of the code. Examples would be branch or path testing

hypermedia (hypertext)

Information structure consisting of nodes of content including link anchors to other nodes of content

iconic models

Graphical models in Euclidean space of artifacts that will occupy Euclidian space when constructed

ility

A quality attribute, or non-functional requirement. The name comes from the form of many requirements such as scalability, reusability, modifiability, etc.

implementation

The phase in the software life-cycle where the actual software is implemented; the result of this phase consists of source code and its documentation [adapted from Stahl and Volter 2004]

ill-structured problem

Term used by Reitman and later by Simon to refer to open-ended problems, like software design, that cannot be uniquely decomposed into verifiable steps. See also "wicked problems"

inconsistency

A state in which two or more overlapping elements of different software models make assertions about aspects of the system they describe which are not jointly satisfiable [Spanoudakis and Zisman 2001]

incremental delivery

A software development process where the software is developed and delivered in increments rather than as one completed system at the end of development

inspection

A verification technique that involves reviewing the software artifacts to look for defects

integrated rationale

Rationale for a software system that is stored with or as part of the software that it describes and explains

integration testing

Tests performed to ensure that the subsystems that comprise a software system work correctly together

interaction

The mutual influence of two actors, and/or components. Interaction is performed via an interface [Chaudron et al. 2004]

interface

For two components, or a component and an external actor, a model of types of the messages that are exchanged and the order in which this may occur [Chaudron et al. 2004]

Issue Based Information System (IBIS)

Issue-based Information System is a way of modeling argumentation; it has been invented by Rittel in 1970. See also "wicked problems"

Knowing-in-Action

In Schön's theory of Reflective Practice, the process of performing tasks in an intuitive, non-reflective manner that involves unselfconscious engagement in the task at hand

Lehman's laws

Eight laws that describe how software systems evolve

maintenance

Software modifications made to systems after they have been delivered to the customer

metaphor (in software design)

A direct comparison of a software component to a physical object (desk top), a social institution (library), an animate entity (garbage collector), etc. to assist in comprehension and communication. See also "anthropomorphism"

metrics

Measurements of software properties or processes used to evaluate the software and/or the process by which it was developed

model

A formal representation of an aspect of a system. Typical examples are data models that give a static view and process models that give a dynamic view [Chaudron et al. 2004]

model (with concurrent multiple views of 'RBSE')

Describes logical organization, dynamic behavior, software organization, process decomposition, and physical realization

Model Driven Architecture (MDA)

In MDA, models are the central elements of the software development process. The main goal is to transform platform-specific models, possibly automatically, into platform-independent models [Kleppe et al. 2003]

model driven development

Software is developed by first building a model of the system and then transforming that into the code

multi-scale visualization

Visualizations with qualitatively/structurally distinct levels of zoom

narrative

An informal design rationale representation in which stories or scenarios describe how and why a design decision was reached, or how and why a user experienced a design system

naturalistic decision making

Decision making methodology that emphasizes identifying and leveraging the strengths of human decision making, instead of merely remediating weaknesses and fallacies

non-functional requirement (NFR)

Software requirements that describe desirable properties of the software that do not map to specific functionality but instead apply to the system as a whole

ontology

A set of entities, their definitions, and the relationships between them

open source software

Software where the source code is freely available for use and modification

open-closed principle

A class that follows the open-closed principle is open to extension and closed to modification

operational environment

The environment in which the software is operating after delivery

pareto optimality

A solution where it can not be improved further by one criteria without worsening in another

pattern-mining

A process of extracting and documenting architecturally significant infor-mation from patterns to support the architecture design and evaluation process

perfective maintenance

Changes to improve a software system that are not in response to defects

Personal Software Process (PSP)

A methodology for improving individual software process by collecting and using metrics captured during software development on an individual basis

platform

A set of subsystems and technologies that provide a coherent set of functionality through interfaces and specified usage patterns, which any application supported by that platform can use without concern for the details of how the functionality provided by the platform is implemented [Kleppe et al. 2003]

portability

Portability is the property of a system which permits it to be mapped from one environment to a different environment [Randal and Buxton 1970]

post-specification traceability

The ability to trace from a software requirement forward to the code that implements it and the tests that verify it has been implemented

Potts-and-Bruns rationale approach

A modification of IBIS for use in software design. The crucial innovation of their approach is to include in their schema elements that represented "intermediate artifacts," i.e. the various models, documents and prototypes produced during design to represent the software being designed

pre-specification traceability

The ability to trace from a software requirement backward to the customer request that it responds to

preventative maintenance

Changes to a software system to avoid anticipated future problems

problem-based evaluation

Informal design rationale approach in which a set of problem scenarios are used to analytically evaluate a design proposal

Procedural Hierarchy of Issues (PHI)

A refinement of IBIS whose main innovation is to show that frequently the decision on one issue depends on the decisions made on others. PHI models rationale as a quasi-hierarchical structure of issues linked by dependency relationships

product line

A collection of existing and potential products that address a coherent business area and share a set of similar characteristics. All these products are made by the same process and for the same purpose, and differ only in style, model or size [http://w3.umh.ac.be/genlog/SE/SE-contents.html]

program comprehension

The process of understanding the source code of a software system

quality assurance

An approach to ensure that the software product, and the processes used to develop it, conform to the software specification and other required standards and procedures

Questions-Options-Criteria (QOC)

A rationale approach resembling IBIS but not derived from it. Like IBIS, QOC centers on decision tasks that are represented as *questions*. But unlike IBIS, QOC deals only with "design space" questions, i.e. those that determine features of the designed artifact, rather than the wider range of questions dealt with by IBIS. QOC's main innovation is the use of criterion-based evaluation in the first level of argumentation of decision alternatives (options)

Rapid Application Development (RAD)

A software development process that makes heavy use of Computer Aided Software Engineering (CASE) tools to quickly build software systems

rationale approach

A way of modeling and using rationale

rationale database

Structured repository of reusable rationales, accessible via type of system, application, scenario, issue, position, argument, etc.

Rationale Based Software Engineering (RBSE)

The research on and use of rationale capture and delivery to support every aspect of software engineering.

rationale capture problem

The difficulty of capturing rationale in a structured form. This is considered by many to be the main impediment to widespread use of rationale approaches in artifact creation in general and software development in particular

rationale management

Is concerned with capturing, representation, retrieval, and using the reasoning behind decisions made during the system development process

rationale management system

Software tools developed to support rationale management

RATspeak

Burge's extension of DRL to make it more suitable for software engineering. RATspeak introduces new types of elements into its schema and provides an argument ontology tailored to software engineering. These additions enable automated checking and inference-making

recognition-primed decision model

Expert decision model in which situations are rapidly classified and addressed as exemplars of known prototypes

re-engineering

Re-writing all or part of an existing software system to improve its quality. This typically refers to a legacy system currently in use that is no longer maintainable.

refactoring

Making modifications to code to correct "bad smells" and to prepare the code for future extension. Refactoring does not add or change functionality

Reflection-in-Action

In Schön's theory of Reflective Practice, the process of explicitly reflecting on why an intuitive performance of a task *broke down*, i.e. led to unforeseen results

Reflective Practice

Schön's theory that design and other practical problem solving activities consist of repeated alternation between two processes that he labeled Knowing-in-Action and Reflection-in-Action

regression testing

Repeating earlier tests on a previously tested product to ensure that new modifications have not introduced defects into existing code

requirement

A property that is demanded to be fulfilled by a software system [adapted from Chaudron et al. 2004]

requirements elicitation

Obtaining requirements from various system stakeholders by a variety of techniques including interviews, observation, and prototyping

requirements engineering

The process of eliciting and documenting software requirements to ensure completeness and consistency

requirements traceability

The ability to trace the impact of a requirement on the delivered system in order to ensure that all requirements have been satisfied

re-use

Using existing code when building a new system. This may or may not involve modifying that code

reverse engineering

Using the source code to create the specification and models that describe the system

satisficing

Evaluation metric used in problem solving and decision making in which the first acceptable solution is adopted. Contrasts with optimization

Scenario Claims Analysis (SCA)

Informal design rationale in which core user interactions afford by a software system are described by scenarios and implicit design tradeoffs in the scenarios (claims)

semantic inference

Inferences that use the semantics of the items being analyzed. In the case of rationale, it means inferencing over the contents and not just the structure.

service oriented development

Systems are built around a Service Oriented Architecture (SOA) that builds a system using loosely coupled distributed services where services can be accessed transparent of their platform implementation

situated cognition

Approach to analyzing human thought that regards the actors and objects of social and material contexts as constitutive resources

social capital

A sense of generalized reciprocity within a social group

social loafing

The tendency of people to work less hard when working in the context of others doing the same work

Software Engineering (SE)

Software engineering is the development and maintenance of software by the systematic application of engineering techniques in the software domain [adapted from IEEE Std 610.12-1990]

Software Engineering Institute (SEI)

A federally funded (USA) software engineering research center that conducts software engineering research in a number of areas that include software architecture, software product lines, and software process improvement

Software Engineering Rationale (SER)

SER emphasizes that rationale models are used during all activities of software development, including requirements engineering, architectural design, implementation, testing, and system deployment

Software Process (SP)

A related set of activities and processes that are involved in developing and evolving a software system [Sommerville 2007]

software process improvement

Evaluating and modifying a software development process to achieve a higher level of repeatability, maturity, and performance

solution-first bias

Tendency of designers to rapidly frame a solution to a problem they do not yet fully understand

Spiral model

A software lifecycle model that utilizes iteration where each trip around the spiral involves determining objectives, assessing risk, developing the current phase of the product, and planning the next phase

stakeholder

Anyone who has interest in the success of the software project

syntactic inference

Inference over rationale that looks only at the structure of the argumentation and not at the contents

system

A generic term for a group of interrelated, interdependent or interacting elements serving a collective purpose [Chaudron et al. 2004]

system architecture

The fundamental organization of a system, embodied in its components, their relationships to each other and the environment, and the principles governing its design and evolution [IEEE Std 610.12-1990]

system engineering

The process of developing a system that must fulfill a certain purpose using the systematic application of engineering techniques, and of which software engineering is a part, provided the system has a software subsystem [Chaudron et al. 2004]

team software process

A methodology for working in teams that includes a framework for managing, tracking, and reporting on the team's performance

test case

A software testing document that consists of the input for and expected result of running the test

test driven development

A software development methodology where unit tests are written first and then the code is written to pass the test

traceability

The degree to which a relationship can be established between two or more products of the development process [adapted from IEEE Std 610.12-1990]

testing

Executing a piece of software to look for defects

traditional approach to rationale capture

The approach in which rationale is structured according a given rationale schema as it is recorded

Unified Modeling Language (UML)

A standardized specification language for object modeling [http://www.omg.org/uml]

Unified Process (UP)

A software development framework utilizing incremental and iterative development. The Unified Process contains four phases: inception, elaboration, construction and transition

unit testing

Testing the smallest testable pieces of source code

usage-centric rationale approaches

Rationale approaches that deal with rationale derived from the experiences of users as they use artifacts

validation

Ensuring that the software system conforms to its specification

Value-based Software Engineering

A theory of software development where the emphasis is on providing value to all the system stakeholders

verification

Ensures that the software system is fit for its intended use

view

A view is a representation of a whole system from the perspective of a related set of concerns [IEEE Std 1471-2000]

viewpoint

A viewpoint is a specification of the conventions for constructing and using a view. Typical viewpoints are structure, behavior, functionality, security, distribution, performance, usability, usefulness, and reliability [IEEE Std 1471-2000]

V-model

A software development lifecycle model where each development stage is paired with the corresponding verification stage

war-room (in design)

A dedicated design workroom in which analyzes and artifacts are pinned to the walls

Waterfall model

A sequential software development model where development flows from one stage to the next

white-box Testing

Software testing that is based on information about the structure of the code. Examples would be branch or path testing.

wicked problems

Rittel's theory of problems of artifact creation as fundamentally open-ended and potentially controversial. According to Rittel, such problems cannot be solved using a strictly scientific approach or purely automated methods. Instead, their solution requires methods that support creative human problem solving by means of an "argumentative approach." Wicked problems theory was used to justify Rittel's pioneering work on design rationale

Index